# Discrete
# Mathematics

Other titles in this series

**Linear Algebra**
R B J T Allenby

**Mathematical Modelling**
J Berry and K Houston

**Particle Mechanics**
C Collinson and T Roper

**Vectors in 2 or 3 Dimensions**
A E Hirst

**Numbers, Sequences and Series**
K E Hirst

**Groups**
C R Jordan and D A Jordan

**Probability**
J McColl

*In preparation*

**Ordinary Differential Equations**
W Cox

**Calculus and ODEs**
D Pearson

**Analysis**
E Kopp

**Statistics**
A Mayer and A M Sykes

Modular Mathematics Series

# Discrete Mathematics

## A Chetwynd and P Diggle

*Department of Mathematics and Statistics*
*Lancaster University*

OXFORD  AUCKLAND  BOSTON  JOHANNESBURG  MELBOURNE  NEW DELHI

Butterworth-Heinemann
Linacre House, Jordan Hill, Oxford OX2 8DP
225 Wildwood Avenue, Woburn, MA 01801-2041
A division of Reed Educational and Professional Publishing Ltd

$\mathcal{R}$ A member of the Reed Elsevier plc group

First published by Arnold 1995
Reprinted by Butterworth-Heinemann 2001

**British Library Cataloguing in Publication Data**
A catalogue record for this book is available from the British Library

ISBN 0 340 61047 6

Typeset in 10/12 Times by
Paston Press Ltd, Loddon, Norfolk
Printed and bound in Great Britain by
J W Arrowsmith Ltd, Bristol

PLANT A TREE
*BTCV*
*British Trust for*
*Conservation Volunteers*

FOR EVERY TITLE THAT WE PUBLISH, BUTTERWORTH-HEINEMANN
WILL PAY FOR BTCV TO PLANT AND CARE FOR A TREE.

# Contents

# Series Preface

This series is designed particularly, but not exclusively, for students reading degree programmes based on semester-long modules. Each text will cover the essential core of an area of mathematics and lay the foundation for further study in that area. Some texts may include more material than can be comfortably covered in a single module, the intention there being that the topics to be studied can be selected to meet the needs of the student. Historical contexts, real life situations, and linkages with other areas of mathematics and more advanced topics are included. Traditional worked examples and exercises are augmented by more open-ended exercises and tutorial problems suitable for group work or self-study. Where appropriate, the use of computer packages is encouraged. The first level texts assume only the A-level core curriculum.

Professor Chris D. Collinson
Dr Johnston Anderson
Mr Peter Holmes

# Preface

In this book, our aim has been to provide an introduction to some fundamental concepts in modern mathematics (logic; sets; relations and functions), and to develop these ideas in the context of three particular topics: combinatorics (the mathematics of counting); probability (the mathematics of chance); graph theory (the mathematics of connections in networks).

Throughout, we have made extensive use of worked examples to develop general ideas, as it has been our experience that for many students this is a more effective learning method than the traditional format of theorem–proof–example. Many of these worked examples are deliberately straightforward, but some are quite challenging. This applies also to the student exercises at the end of each section.

We have tried to write at a level suitable for the beginning of a first-year undergraduate course in mathematics, and most of the material in the book appears in first-year courses taught at Lancaster University. On the basis of our experience at Lancaster, we would suggest that Chapters 1 to 3, followed by Chapters 4 and 5 *or* Chapter 6, contain about the right amount of material for a 20-lecture module at the beginning of the first year. Alternatively, Chapters 1 to 4 would be appropriate for a slightly shorter (or more leisurely!) introductory module, in which case each of Chapters 5 and 6 could be used as springboards for more ambitious modules in probability and graph theory, respectively.

Discrete mathematics is of interest in its own right but we have found that it also provides an intuitively natural introduction to formal mathematical concepts which are of wider applicability—for example, the ideas underlying different types of proof, and the definition of a function. Discrete mathematics also has important applications in other branches of science—in particular, formal methods in computer science are largely based on discrete mathematics.

Most of the work on the book was done at Lancaster, and we are grateful to our colleagues Anthony Ledford, who gave us valuable comments on a first draft, and Barry Rowlingson, who helped us with computing. We also thank the Department of Statistics, University of Adelaide for their hospitality during March 1994. Finally, we thank our respective partners for their understanding of the long hours we spent working on the book when we could otherwise have been at home.

Dr Amanda Chetwynd
Professor Peter Diggle
Lancaster University
April 1995

# 1 • Logic

The Greek philosopher and scientist Aristotle (384–322 BC) is said to be the first person to have studied logical reasoning.

**Fig 1.1** Aristotle (Brown Brothers)

The *Oxford Illustrated Dictionary* defines the word 'logical' to mean correctly reasoned.

In mathematics we interpret this definition in its strict sense, to mean that each step in an argument must follow unambiguously from the preceding step, with no room for subjective judgement. In this way, we try to establish whether or not formal statements are true, and to demonstrate interesting consequences. Logical reasoning is the essence of mathematics, and logic is therefore a natural starting point for our study of mathematics.

In this chapter, we introduce you to the basic language and notation of formal logic, and study *truth tables*, which provide a method for finding out whether a complicated statement is true or false by breaking the statement down into a collection of simpler statements whose truth or otherwise is obvious. We then discuss the idea of a mathematical *proof*, and show you how to carry out several different types of proof: direct proof, contrapositive proof, proof by contradiction, proof by counterexample and finally proof by mathematical induction.

## 1.1 Introduction

The building block of formal logic is the *proposition*. A **proposition** is any formal statement which is either true or false, but is not a matter of opinion.

## Example 1

Consider the four propositions

- a mouse is a small cat
- Jenny is a pilot
- there is life on Mars
- 37 is a prime number

Only the last of these is obviously mathematical, but the key feature of each of them, and of any proposition, is that it may either be true or false (although we may not know which!), but it is not a matter of opinion.

We can abbreviate our propositions and say that $p$ stands for 'a mouse is a small cat', $q$ stands for 'Jenny is a pilot', $r$ stands for 'there is life on Mars' and $s$ stands for '37 is a prime number.' The *truth value* of a proposition is just a statement of whether it is true or false. We write $T$ for 'true' and $F$ for 'false'.

## Example 2

An example of a true proposition is

- 37 is a prime number.

An example of a false proposition is

- a mouse is a small cat.

An example of a proposition whose truth value we do not know is

- there is life on Mars.

A simple proposition is usually called a **primitive proposition**. We can *negate* a simple proposition, i.e. make the opposite proposition and thereby reverse its truth value, or we can combine primitive propositions in different ways, using logical operators like **not**, **or**, **and**, and so obtain **compound propositions**. In the rest of this chapter, whenever an English language word like 'not' is being used as a logical operator, we will set it in **bold-face** type, to emphasise that we are using the word in its technical sense.

### Negation

'A mouse is not a small cat' is a negation of proposition $p$. To indicate the negation of a proposition $p$ we write **not** $p$; some books use the symbol $\sim p$.

## Example 3

If $q$ is the proposition 'Jenny is a pilot', then the negation of $q$, **not** $q$, is the proposition 'Jenny is not a pilot'.

## Example 4

If $r$ is the proposition 'there is life on Mars' then the proposition 'there is no life on Mars' is equivalent to **not** $r$.

## Conjunction and disjunction

The statement

> a mouse is a small cat and Jenny is a pilot

is called a **conjunction** of two propositions, whereas the statement

> a mouse is a small cat or there is life on Mars

is a **disjunction** of propositions.

The conjunction of $p$ and $q$ is denoted by $p$ **and** $q$. The disjunction of $p$ or $q$ is denoted by $p$ **or** $q$. Some books use & or $\wedge$ instead of **and** and $\vee$ instead of **or**.

Consider the following compound propositions:

- 37 is a prime number and Jenny is a pilot
- 37 is a prime number or Jenny is a pilot

Suppose Jenny is not, in fact, a pilot. However, 37 is a prime number. Is each compound proposition true? The rules for conjunction and disjunction are:

- for $p$ **and** $q$ to be true both $p$ must be true and $q$ must be true
- for $p$ **or** $q$ to be true either $p$ must be true or $q$ must be true, or both must be true

So in this case, the conjunction is false but the disjunction is true.

## ● *Example 5*

Let $p$ be the proposition that 'the world is overpopulated' and $q$ be the proposition that 'the population of the world can fit onto the Isle of Wight'. Describe the following propositions in English:

1. $p$ **and** $q$
2. $p$ **or** $q$
3. **not** $q$

SOLUTION
1. The world is overpopulated and the population of the world can fit onto the Isle of Wight.
2. Either the world is overpopulated or the population of the world can fit onto the Isle of Wight.
3. The population of the world cannot fit onto the Isle of Wight.

## EXERCISES ON 1.1

1. Let $p$ be the proposition that 'the number of students is increasing' and $q$ be the proposition that 'the student grant is decreasing'. Describe the following propositions in English:
   (a) $p$ **and** $q$
   (b) $p$ **or** $q$
   (c) **not** $q$
2. Abbreviate the following two propositions to symbols:
   Anne is a mathematician.
   Anne won the race.

Now express each of the following sentences as compound propositions using logic symbols:
(a)  Anne is a mathematician and she won the race.
(b)  Anne is not a mathematician.
(c)  Either Anne is a mathematician or she won the race.
(d)  Anne is a mathematician and she did not win the race.
(e)  Anne is not a mathematician, nor did she win the race.

3.   Abbreviate the following three propositions to symbols:
The students found the material too difficult.
The students thought the lecturer's presentation was good.
The students enjoyed the course.
Now express each of the following sentences as compound propositions using logic symbols:
(a)  The students found the material too difficult and thought the lecturer's presentation was good.
(b)  The students enjoyed the course but thought the lecturer's presentation was not good.
(c)  The students either found the material too difficult or thought the lecturer's presentation was good, but not both.
(d)  The students did not enjoy the course and they found the material too difficult.
(e)  The students found the material too difficult, thought the lecturer's presentation was not good and did not enjoy the course.

4.   For two propositions $p$ and $q$, is the compound proposition

($p$ **and** $q$) **and not** ($p$ **or** $q$)

(a)  always true?
(b)  always false?
(c)  possibly true or false, depending on the truth value of $p$ and the truth value of $q$?

5.   For two propositions $p$ and $q$, is the compound proposition

($p$ **or** $q$) **or not** ($p$ **and** $q$)

(a)  always true?
(b)  always false?
(c)  possibly true or false, depending on the truth value of $p$ and the truth value of $q$?

## 1.2 Truth tables

The logical operator **not** has the effect of reversing the truth value of any proposition to which it is applied. We can efficiently summarise the effect of logical operators using a **truth table**.

To illustrate the idea of a truth table, we will first construct the table for **not** $p$. Since $p$ is a proposition, its possible truth values are true ($T$) and false ($F$). The truth table for **not** $p$ has two rows to correspond to the two possible truth values of the proposition $p$:

| $p$ | **not** $p$ |
|-----|-------------|
| $T$ | $F$ |
| $F$ | $T$ |

Now let us construct the truth table for a compound proposition.

To give the truth tables for **and** and for **or** we need to start with two propositions, say $p$ and $q$. There are now four possibilities for assigning pairs of truth values ($T$ or $F$) to the pair of propositions $p$ and $q$. Hence, each table has four rows, as follows:

| $p$ | $q$ | $p$ **and** $q$ |
|-----|-----|-----------------|
| $T$ | $T$ | $T$ |
| $T$ | $F$ | $F$ |
| $F$ | $T$ | $F$ |
| $F$ | $F$ | $F$ |

| $p$ | $q$ | $p$ **or** $q$ |
|-----|-----|----------------|
| $T$ | $T$ | $T$ |
| $T$ | $F$ | $T$ |
| $F$ | $T$ | $T$ |
| $F$ | $F$ | $F$ |

Truth tables become very useful when we want to establish the truth value for a complicated compound proposition. This is because the truth table allows us to break down the complete, complicated proposition into simpler pieces.

## Example 6

Give the truth table for the compound proposition (**not** $p$) **or** $q$.

SOLUTION

| $p$ | $q$ | **not** $p$ | (**not** $p$) **or** $q$ |
|-----|-----|-------------|--------------------------|
| $T$ | $T$ | $F$ | $T$ |
| $T$ | $F$ | $F$ | $F$ |
| $F$ | $T$ | $T$ | $T$ |
| $F$ | $F$ | $T$ | $T$ |

## Example 7

Give the truth table for the compound proposition **not** ($p$ **or** $q$).

SOLUTION

| $p$ | $q$ | $p$ **or** $q$ | **not** ($p$ **or** $q$) |
|-----|-----|----------------|--------------------------|
| $T$ | $T$ | $T$ | $F$ |
| $T$ | $F$ | $T$ | $F$ |
| $F$ | $T$ | $T$ | $F$ |
| $F$ | $F$ | $F$ | $T$ |

In these two examples, notice how we used brackets to make the statement of each compound proposition unambiguous. Now let's try a more complicated sentence.

## Example 8

What is the truth value of the following compound proposition?

> Either spaghetti grows on trees and Margaret Thatcher was the first British female Prime Minister, or it is not true that the dodo is extinct.

SOLUTION

Let $p$ stand for the proposition 'spaghetti grows on trees', $q$ stand for the proposition 'Margaret Thatcher was the first British female Prime Minister' and $r$ stand for the proposition 'the dodo is extinct'.

Then our proposition reads $(p$ **and** $q)$ **or** (**not** $r)$, and the truth values of the three primitive propositions $p$, $q$ and $r$ are, respectively, $F$, $T$ and $T$.

If we replace each symbol by its truth value, we can simplify progressively by using the truth tables from our earlier examples, to get

$$(F \text{ and } T) \quad \text{or} \quad \text{not } T$$
$$F \qquad \text{or} \qquad F$$
$$F$$

So the compound proposition is false.

## Example 9

By using truth tables, show that

> **not** $(p$ **and** $q)$ is equivalent to (**not** $p)$ **or** (**not** $q)$.

SOLUTION

| $p$ | $q$ | $p$ and $q$ | not($p$ and $q$) | not $p$ | not $q$ | (not $p$) or (not $q$) |
|-----|-----|-------------|------------------|---------|---------|------------------------|
| $T$ | $T$ | $T$ | $F$ | $F$ | $F$ | $F$ |
| $T$ | $F$ | $F$ | $T$ | $F$ | $T$ | $T$ |
| $F$ | $T$ | $F$ | $T$ | $T$ | $F$ | $T$ |
| $F$ | $F$ | $F$ | $T$ | $T$ | $T$ | $T$ |

Notice that the truth values in the two columns headed **not** $(p$ **and** $q)$ and (**not** $p)$ **or** (**not** $q)$ are identical.

## Example 10

Use truth tables to prove that

> $p$ **and** $(q$ **or** $r)$ is equivalent to $(p$ **and** $q)$ **or** $(p$ **and** $r)$.

SOLUTION

| $p$ | $q$ | $r$ | $(q$ or $r)$ | $p$ and $(q$ or $r)$ | $p$ and $q$ | $p$ and $r$ | $(p$ and $q)$ or $(p$ and $r)$ |
|---|---|---|---|---|---|---|---|
| $T$ | $T$ | $T$ | $T$ | $T$ | $T$ | $T$ | $T$ |
| $T$ | $T$ | $F$ | $T$ | $T$ | $T$ | $F$ | $T$ |
| $T$ | $F$ | $T$ | $T$ | $T$ | $F$ | $T$ | $T$ |
| $T$ | $F$ | $F$ | $F$ | $F$ | $F$ | $F$ | $F$ |
| $F$ | $T$ | $T$ | $T$ | $F$ | $F$ | $F$ | $F$ |
| $F$ | $T$ | $F$ | $T$ | $F$ | $F$ | $F$ | $F$ |
| $F$ | $F$ | $T$ | $T$ | $F$ | $F$ | $F$ | $F$ |
| $F$ | $F$ | $F$ | $F$ | $F$ | $F$ | $F$ | $F$ |

Since the two columns, headed $p$ **and** $(q$ **or** $r)$ and $(p$ **and** $q)$ **or** $(p$ **and** $r)$, of the truth tables agree, the two compound propositions are identical. This proves that $p$ **and** $(q$ **or** $r)$ is equivalent to $(p$ **and** $q)$ **or** $(p$ **and** $r)$.

## ● *Example II*

Propositions like

● the lecturer is either a woman or a man
● people either like moths or they don't

are always true, and are called **tautologies**.
   Propositions like

● Marcia is a girl and is a boy
● $x$ is prime and $x$ is an even integer greater than 6

are always false, and are called **contradictions**.

## EXERCISES ON 1.2

1.  Give the truth table for **not** (**not** $p$).
2.  Use truth tables to show that, for any proposition $p$,
    (a)  $p$ **or** (**not** $p$) is a tautology
    (b)  $p$ **and** (**not** $p$) is a contradiction
    (c)  $p$ **and not** $(p$ **or** $q)$ is a contradiction
3.  Find the primitive propositions in the following sentence, write the sentence in logic notation and evaluate its truth value:

    Italy is not a member of the European Union, which is a community of nations whose parliament meets in Brussels.

4.  Show that **not** $(p$ **or** $q)$ is equivalent to (**not** $p$) **and** (**not** $q$).
5.  Which of the following propositions are tautologies, contradictions or neither?
    (a)  The world is round and life is sweet.
    (b)  All men are liars and all men tell the truth.
    (c)  All even numbers are divisible by 2 and all numbers that are not odd are divisible by 2.

# .3 Conditional propositions

n this section we shall look at **conditional propositions**. Examples include:

- if tomorrow is Saturday then today is Friday,
- if $x$ is a positive integer then $x$ is bigger than 0.

To attach a truth value to a conditional proposition, we have to draw a very careful distinction between logical truth and factual truth. A conditional proposition of the form

> if $p$ is true, then $q$ follows

or in more compact mathematical notation

> if $p$ then $q$

is logically correct (and therefore 'true' in a logical sense) if the truth of $p$ implies that $q$ must also be true. If we start from a false premise (i.e. if $p$ is false), we may reach either a correct conclusion ($q$ is true) or an incorrect conclusion ($q$ is false), by a logically correct argument as the following two examples show.

## ● *Example 12*

If all integers are even, then 7 is divisible by 2.

    This conditional proposition is logically correct. If all integers are even, then all integers are divisible by 2, and therefore 7 must be divisible by 2 because 7 is an integer.

## ● *Example 13*

If all integers are even, then 8 is divisible by 2.

    This conditional proposition is also logically correct. If all integers are even, then all integers are divisible by 2, and therefore 8 must be divisible by 2 because 8 is an integer.

However, if we start from a correct premise ($p$ is true) we cannot reach an incorrect conclusion ($q$ is false) unless we make a logical error in our argument.

    It follows that the truth value of the conditional proposition

> if $p$ then $q$

is 'true' unless $p$ is true and $q$ is false.

## ● *Example 14*

Which of the following propositions are true and which are false?

1. If the earth is round then the earth travels round the sun.
2. If spiders have 10 legs then Rupert has a yellow scarf.
3. If *bonjour* is French then *auf wiedersehen* is Italian.

SOLUTION
1.  True. The first part is true, so we have to check the second part. The second part is also true, so the conditional proposition is true.
2.  True. Because the first part is false (spiders don't have 10 legs), it does not matter what the second part says.
3.  False. Because the first part is true we again have to check the second part, but this is false (*auf wiedersehen* is German!).

Notice that propositions which are logically correct may still be nonsensical in an everyday sense!

## Example 15

Consider the proposition *p*: if I get less than 35% on the test than I shall study harder. Which of the following situations are consistent with *p*?

1.  I get less than 35%, I begin to study harder.
2.  I get less than 35%, I decide there is no point in studying harder.
3.  I get more than 35%, I am pleased so I study harder.
4.  I get more than 35%, I do not study harder.

SOLUTION
If I get less than 35% then *p* says that I shall study harder. So the first example is consistent with *p*, and the second is false. However, *p* says nothing about what happens if I get more than 35%, so the last two statements are both consistent with *p*.

A compact notation for the conditional proposition

if *p* then *q*

is

$$p \Rightarrow q$$

The truth table for $p \Rightarrow q$ is:

| *p* | *q* | $p \Rightarrow q$ |
|-----|-----|-------------------|
| *T* | *T* | *T* |
| *T* | *F* | *F* |
| *F* | *T* | *T* |
| *F* | *F* | *T* |

## If and only if

Suppose you say to one of your friends

if I do well at mathematics then I will go to classes,

whereas to another friend you say

if I go to classes then I will do well at mathematics.

These are very different propositions! What would it mean to say that both propositions were true? The conjunction of the two is the proposition that

I do well at mathematics if and only if I go to classes.

## ● *Example 16*

Consider the proposition 'I go to the supermarket if and only if I have run out of food'. This is the same as saying: 'If I have run out of food then I go to the supermarket, and if I go to the supermarket then I have run out of food'.

The symbol for 'if and only if' is ⇔. Here is the truth table for $p \Leftrightarrow q$:

| $p$ | $q$ | $p \Leftrightarrow q$ |
|-----|-----|-----------------------|
| $T$ | $T$ | $T$ |
| $T$ | $F$ | $F$ |
| $F$ | $T$ | $F$ |
| $F$ | $F$ | $T$ |

## ● *Example 17*

We can prove that $p \Leftrightarrow q$ is equivalent to

$$p \Rightarrow q \text{ and } q \Rightarrow p$$

using truth tables, as follows:

| $p$ | $q$ | $p \Rightarrow q$ | $q \Rightarrow p$ | $(p \Rightarrow q)$ **and** $(q \Rightarrow p)$ | $p \Leftrightarrow q$ |
|-----|-----|-------------------|-------------------|------------------------------------------------|-----------------------|
| $T$ | $T$ | $T$ | $T$ | $T$ | $T$ |
| $T$ | $F$ | $F$ | $T$ | $F$ | $F$ |
| $F$ | $T$ | $T$ | $F$ | $F$ | $F$ |
| $F$ | $F$ | $T$ | $T$ | $T$ | $T$ |

The truth values in the two columns headed $(p \Rightarrow q)$ **and** $(q \Rightarrow p)$ and $p \Leftrightarrow q$ are identical, so the corresponding compound propositions are equivalent.

The connective ⇒ can be expressed in several different ways:

- $p \Rightarrow q$
- $p$ is **sufficient** for $q$
- if $p$ then $q$
- $p$ implies $q$
- $p$ only if $q$

This can be confusing! But it gets less so with experience.

We can also use the implies arrow in the opposite direction to mean 'is implied by'. Thus, '$p$ implies $q$' is the same as '$q$ is implied by $p$', that is

$$p \Rightarrow q \text{ is the same as } q \Leftarrow p$$

Hence,

- $p \Leftarrow q$
- $p$ is **necessary** for $q$

- if $q$ then $p$
- $p$ is implied by $q$
- $p$ if $q$

We then have similar combined expressions for 'if and only if':

- $p \Leftrightarrow q$
- $p$ is necessary and sufficient for $q$
- $p$ if and only if $q$
- $p$ iff $q$

Notice that $p \Rightarrow q$ is not the same as $q \Rightarrow p$. However, the following truth table shows that $p \Rightarrow q$ is the same as **not** $q \Rightarrow$ **not** $p$. We shall make use of this result in the section on proof later in this chapter.

| $p$ | $q$ | **not** $p$ | **not** $q$ | **not** $q \Rightarrow$ **not** $p$ | $p \Rightarrow q$ |
|-----|-----|-----|-----|-----|-----|
| $T$ | $T$ | $F$ | $F$ | $T$ | $T$ |
| $T$ | $F$ | $F$ | $T$ | $F$ | $F$ |
| $F$ | $T$ | $T$ | $F$ | $T$ | $T$ |
| $F$ | $F$ | $T$ | $T$ | $T$ | $T$ |

## EXERCISES ON 1.3

1. Construct the truth tables for the following statements:
   (a) $p \Leftrightarrow (q \text{ or } p)$
   (b) $p \Leftrightarrow (q \Rightarrow p)$
   (c) $p \Rightarrow (q \Rightarrow r)$
   (d) $p$ **and** **not** $(q \Rightarrow r)$
2. Show that $p \Rightarrow q$ is equivalent to (**not** $p$) **or** $q$.
3. Express $p \Leftrightarrow q$ using only the symbols **not, and, or**.
4. Consider the statement 'Ann is older than Betty unless Betty is younger than Claire' in the following cases:
   (a) Ann is 10, Betty is 11, Claire is 13.
   (b) Ann is 10, Betty is 9, Claire is 14.
   (c) Ann is 10, Betty is 6, Claire is 5.
   (d) Ann is 10, Betty is 11, Claire is 9.
   Draw up a truth table for 'unless'.
5. Express each of the following propositions in logical symbols and decide in each case if the subsequent statement is a valid consequence of the proposition.
   (a) *Proposition:* If I move house, then I will not be able to go on holiday in July.
   *Statement:* Since I am going on holiday in July I will not move house.
   (b) *Proposition:* Either the professor or the student committed the crime. If the professor did it, he would not have been able to give the lecture at 11.00.
   *Statement:* Since the professor gave the lecture at 11.00, the student must have committed the crime.
   (c) *Proposition:* If I get the job and work hard, then I will be promoted.
   *Statement:* I was promoted, so I must have got the job.

(d) *Proposition:* If I get the job and work hard, then I will be promoted.
*Statement:* I was not promoted. Thus, either I did not get the job or I did not work hard.

(e) *Proposition:* If the prize giving runs late, then they may delay the start of the dinner. If the dinner runs late, then the dance will start after 11.00.
*Statement:* The dance started at 11.15, thus the prize giving ran late.

(f) *Proposition:* Either I will get a first in this course, or I will stop drinking. If I stop drinking, I will lose weight.
*Statement:* I got a first, thus I did not lose weight.

# 1.4 Quantifiers

In the propositions that we have seen so far, each statement has been about a particular object. In this section we shall see how to write propositions that are about whole classes of objects.

## ● *Example 18*

Consider the following propositions:

1. All triangles have the sum of their angles equal to 180°.
2. All elephants have trunks.
3. All students like bananas.
4. All odd prime numbers are bigger than 2.
5. There is an integer between 3 and 7 inclusive.
6. There is a cat without a tail.
7. There is a country with no sea boundary.

The English language words 'all' and 'there is' correspond to logical *quantifiers*. We shall use the upside down A symbol, $\forall$, to denote the logical quantifier **for all**, and the back to front E symbol, $\exists$, to denote the logical quantifier **there exists**.

## ● *Example 19*

Rewrite the propositions in Example 18 using the symbols $\forall$ and $\exists$.

SOLUTION
1. ($\forall$ triangles) (sum of the angles of the triangle $= 180°$).
2. ($\forall$ elephants) (the elephant has a trunk).
3. ($\forall$ students) (the student likes bananas).
4. ($\forall$ odd prime numbers) (the odd prime number is bigger than 2).
5. ($\exists$ an integer) (the integer is between 3 and 7 inclusive).
6. ($\exists$ a cat) (the cat is without a tail).
7. ($\exists$ a country) (the country has no sea boundary).

We need to be able to negate propositions which include quantifiers. For example, the proposition 'there is a student who does not like bananas' is a negation of the proposition 'all students like bananas'. Similarly, the proposition 'all cats have tails' is a negation of the proposition 'there is a cat without a tail'.

## Example 20

Negate the propositions in Example 18.

SOLUTION
1. There is a triangle whose sum of angles $\neq 180°$.
2. There is an elephant without a trunk.
3. There is a student who does not like bananas.
4. There is an odd prime number bigger than 2.
5. All integers are not between 3 and 7 inclusive.
6. All cats have tails.
7. All countries have a sea boundary.

We can see that in each negation, 'all' becomes 'there is' and 'there is' becomes 'all'. Also, the statement itself is negated. We now see how this works in a more mathematical setting.

## Example 21

Negate the proposition

All integers are greater than 5.

SOLUTION
We can write this proposition as

$(\forall \text{ integers } x)(x > 5).$

The negation is

$(\exists \text{ an integer } x)(x \leq 5)$

In the negation, $\forall$ has become $\exists$ and $(x > 5)$ is negated to become $(x \leq 5)$. Thus the negated proposition is

There is an integer less than or equal to 5.

## Example 22

The negation of

$(\exists \text{ an integer } x)(3 \leq x \leq 7)$

is

$(\forall \text{ integers } x)(x < 3 \textbf{ or } x > 7).$

*Be careful* – the order of $\forall$ and $\exists$ matters much more than in everyday speech.

Suppose that all the mathematics students on a particular course must share a single copy of the course text-book. In logic notation, we write this as

$(\exists \text{ book})(\forall \text{ students})$

However, if the student grant increases and every student buys their own copy of the text-book, then we write

$(\forall \text{ students})(\exists \text{ book})$

## Example 23

Consider the proposition

Every student has a book which that student reads.

In logic notation, we can write this as

$\forall x \exists B(x \text{ reads } B)$

where $x$ denotes 'student' and $B$ denotes 'book'. We can now negate the proposition, using the rules listed after Example 20, to get

$\exists x \forall B(x \text{ does not read } B)$

Finally, we can translate the negation back into everyday English as

There is a student who does not read any books.

## Example 24

Are the following propositions true or false?

1. $(\forall \text{ integers } x)(x^2 \neq 3)$
2. $(\forall \text{ integers } x)(\exists \text{ an integer } y)(x + y = 2)$
3. $(\exists \text{ an integer } x)(\forall \text{ integers } y)(x + y = 2)$

SOLUTION
1. True—it says that there are no integers whose square is 3.
2. True—it says that for any integer $x$, $2 - x$ is also an integer.
3. False—since $x + y = 2$ is equivalent to $x = 2 - y$, this says that $2 - y$ has the same value, namely $x$, for any integer $y$.

## Example 25

Negate the following proposition:

$(\forall x > 0)(\exists y > 0)(x + y \text{ is prime})$

The negation is

$(\exists x > 0)(\forall y > 0)(x + y \text{ is not prime})$

## EXERCISES ON 1.4

1. Write the following sentences using logic notation, letting $x$ stand for cat and $P(x)$ stand for $x$ likes cream:
   (a) All cats like cream.
   (b) No cats like cream.
   (c) One cat likes cream.
   (d) Every cat hates cream.

2. Write English sentences that are the negations of the following sentences:
   (a) All fish swim.
   (b) Some newspapers exaggerate the truth.
   (c) No students in mathematics are unable to use the computer.
   (d) No presidents are economical with the truth.
3. Translate the following statements into prose:
   (a) $(\forall x > 0)(\exists y > 0)(x^2 = y)$
   (b) $(\forall x)(\forall y)(\exists z)(x + z = y)$
   (c) $\mathbf{not}[(\exists x)(\forall y)(y > x)]$
4. Negate the following statements. Which of the negated statements are true?
   (a) All good students study hard.
   (b) No males give birth to their young.
   (c) All of Shakespeare's plays are comedies.
   (d) $(\forall$ integers $x)(\exists$ an integer $y)(x^2 = y)$
   (e) $(\exists$ an integer $x)(\forall$ integers $y)(x^2 = y)$

# 1.5  Types of proof

An important part of mathematics is proof. There are many different types of proof. In this section we shall look at some of the more common types.

## Direct proof

This is the most straightforward type of proof.

## Example 26

Prove that if $x$ and $y$ are rational numbers then $x + y$ is rational.
   To prove this, we need to remember that rational numbers are numbers which can be expressed as fractions with integer numerator and denominator. For example, $3/7$ is a rational number, but $\sqrt{2}$ is not (as we shall prove in Example 30).

SOLUTION
Since $x$ and $y$ are rational numbers, we can find integers $p, q, m$ and $n$ such that $x = p/q$ and $y = m/n$. Then

$$x + y = \frac{p}{q} + \frac{m}{n} = \frac{pn + mq}{qn}$$

Since $pn + mq$ and $qn$ are both integers, we conclude that $x + y$ is a rational number.

## Example 27

Prove that if $|x| > |y|$ then $x^2 > y^2$.

SOLUTION
Since $|x| > |y|$ then $|x|^2 > |y|^2$. Now $|x|^2 = x^2$ and $|y|^2 = y^2$. Hence $x^2 > y^2$.

## Contrapositive proof

Often when proving an implication, it is helpful to remember that **not** $q \Rightarrow$ **not** $p$ is the same as $p \Rightarrow q$. It may be easier to prove the contrapositive **not** $q \Rightarrow$ **not** $p$ rather than directly proving $p \Rightarrow q$.

## Example 28

Prove that any prime number $n > 2$ must be odd.

SOLUTION

A more formal statement of this result is that for any integer $n > 2$,

  $n$ prime $\Rightarrow n$ odd

Let $p$ be the proposition

  $n$ prime

and $q$ the proposition

  $n$ odd

We are required to prove that $p \Rightarrow q$, which is the same as proving that (**not** $q$) $\Rightarrow$ (**not** $p$). Now, **not** $q$ is the proposition

  $n$ even

and **not** $p$ is the proposition

  $n$  not prime

The contrapositive is therefore

  $n$ even $\Rightarrow n$ not prime

In words, this says that if $n$ is any even number bigger than 2, then $n$ cannot be prime.

  If $n$ is an even number greater than 2, then $n = 2k$ for some value of $k > 1$. Thus, $n$ is divisible by 2 and $n \neq 2$, so $n$ cannot be prime.

## Example 29

Show that for $n$, a natural number, if $n^2 > 25$ then $n > 5$.

SOLUTION

The contrapositive is

  If $n \leq 5$ then $n^2 \leq 25$

If $n \leq 5$ then $n = 1, 2, 3, 4, 5$ and $n^2 = 1, 4, 9, 16, 25$. Hence $n^2 \leq 25$.

## Proof by contradiction (reductio ad absurdum)

This type of proof is often useful when we have to prove something but cannot see how to begin. We assume the opposite of what we are trying to prove and get a crazy result, i.e. a logical contradiction. Hence, our assumption must have been false, and therefore what we were originally required to prove must be true.

  Here is a famous example of a proof by contradiction.

## ● *Example 30*

Prove that $\sqrt{2}$ is not a rational number.

SOLUTION

Suppose that $\sqrt{2}$ is rational. Then, we can find integers $a$ and $b$ such that

$$\sqrt{2} = \frac{a}{b}$$

and we can assume that any common factors in $a$ and $b$ have been cancelled. We now square both sides of the equation to get

$$2 = \frac{a^2}{b^2}$$

Thus

$$2b^2 = a^2$$

Hence $a^2$ is a multiple of 2, and is therefore even. So $a$ must be even, since the square of any odd number is also odd. Thus, 2 divides $a$ and we can write $a = 2x$ for some integer $x$. Hence,

$$2b^2 = (2x)^2$$
$$2b^2 = 4x^2$$
$$b^2 = 2x^2$$

Thus, $b^2$ is even. Hence, $b$ is even. But now $a$ and $b$ have a common factor of 2, which is a contradiction to the statement that $a$ and $b$ have no common factors.

Hence, our initial assumption that $\sqrt{2}$ is rational is false. Thus, $\sqrt{2}$ is *irrational*.

## ● *Example 31*

Prove that if $x^2 - 4 = 0$ then $x \neq 0$.

SOLUTION

Suppose that $x = 0$, then $0^2 - 4 = -4 \neq 0$, which contradicts $x^2 - 4 = 0$. Hence our assumption that $x = 0$ is false and we have proved that $x \neq 0$.

## **Biconditional**

Prove that

$$x^2 + 5x + 6 = 0 \text{ if and only if } x = -2 \text{ or } x = -3$$

SOLUTION

We need to show that

$$x^2 + 5x + 6 = 0 \Rightarrow x = -2 \text{ or } x = -3$$

and

$$x = -2 \text{ or } x = -3 \Rightarrow x^2 + 5x + 6 = 0$$

Assume that $x^2 + 5x + 6 = 0$. Then

$$x^2 + 5x + 6 = 0 \Rightarrow (x + 2)(x + 3) = 0$$
$$\Rightarrow x + 2 = 0 \text{ or } x + 3 = 0$$
$$\Rightarrow x = -2 \text{ or } x = -3$$

Hence, $x^2 + 5x + 6 = 0 \Rightarrow x = -2$ or $x = -3$.
  Assume $x = -2$ or $x = -3$. If $x = -2$ then

$$x^2 + 5x + 6 = (-2)^2 + 5(-2) + 6 = 4 - 10 + 6 = 0$$

and if $x = -3$ then

$$x^2 + 5x + 6 = (-3)^2 + 5(-3) + 6 = 9 - 15 + 6 = 0$$

Hence, $x = -2$ or $x = -3 \Rightarrow x^2 + 5x + 6 = 0$.
  In simple examples like the one above, the two sub-proofs are so similar that it hardly seems necessary to give them both. But it is important to realise that both sub-proofs *are* strictly necessary, because in more complicated examples it is easy to get the two mixed up and reach a wrong conclusion.

## Example 32

Is it true that $\sin(\theta) = \cos(\theta)$ if and only if $\theta = 45°$?

SOLUTION
If $\theta = 45°$, then $\sin(\theta) = 1$ and $\cos(\theta) = 1$, so clearly $\sin(\theta) = \cos(\theta)$. If you only think of 'sin' and 'cos' in terms of the angles of a triangle, you might conclude that $\theta = 45°$ is the only solution to the equation $\sin(\theta) = \cos(\theta)$, and the proposition of this example would indeed be true. However, the mathematical functions $\cos()$ and $\sin()$ are defined for all real numbers $\theta$, and the equation $\sin(\theta) = \cos(\theta)$ is satisfied if $\theta$ is expressible as $\theta = (45 + 180n)°$ for any integer $n$, so the proposition is false.

## Counterexample
### Example 33

Is it true that

$$(\forall \text{ integers } x)(\exists \text{ an integer } y)(x = y^2)$$

that is, for every integer $x$ there is an integer $y$ where $y^2 = x$?

SOLUTION
Take $x = 5$. There is no integer $y$ such that $y^2 = 5$.
  Hence, $(\forall \text{ integers } x)(\exists \text{ an integer } y)(x = y^2)$ is not true.

## Example 34

Show that the statement

  Every number is greater than the sum of its proper divisors

is false.

SOLUTION

The sum of the proper divisors of 12 is $1+2+3+4+6=16$. Hence, 12 is a counterexample.

A *single* counterexample will do to disprove a proposition, but to prove a proposition we need to check *every* possible example.

Consider the statement

$2^n \geq n^2$ for all positive integers $n$.

The statement is false, because $2^3 = 8$ is less than $3^2 = 9$. Suppose we modify the statement to

$2^n \geq n^2$ for all positive integers $n \geq 4$.

We shall try the first few values. Putting $n = 4$, the claim is that $2^4 \geq 4^2$, i.e. $16 \geq 16$, which is true. Putting $n = 5$, we get $2^5 \geq 5^2$, i.e. $32 \geq 25$, again true. If you carry on like this, you may become convinced that the statement is true for every integer $n \geq 4$, but a proof requires you to show the result for all possible values of $n \geq 4$. In fact, this proposition *is* true. Can you prove it? If not, the next section will show you how.

## EXERCISES ON 1.5

1. Prove that $\sqrt{5}$ is not a rational number.
2. Find a counterexample to each of the following statements:
   (a) If $n = p^2 + q^2$ where $p$ and $q$ are primes then $n$ is prime.
   (b) If $a > b$ then $a^2 > b^2$.
   (c) If $x^4 = 1$ then $x = 1$.
3. Give a direct proof that if $x$ and $y$ are odd integers then $x + y$ is even.
4. Give a direct proof that if $x$ and $y$ are odd integers then $xy$ is odd.
5. Prove using the contrapositive that if $x^2 - 4 < 0$ then $-2 < x < 2$.
6. Write down the contrapositive of each of the following statements:
   (a) If I study mathematics then I get a job.
   (b) If $x$ is an integer greater than zero then $x^2$ is non-negative.
   (c) If the earth goes round the sun then the moon goes round the earth.
   (d) If there are a finite number of prime numbers then the product of all the primes plus 1 is not prime.

# 1.6 Mathematical induction

This is another type of proof, but because it is so important we have given it a section of its own.

## Dominoes

Consider the following situation:

I have stood on end a series of dominoes. Each domino is two inches from its neighbour and in front of the first domino is a pea shooter. I fire the pea shooter at the first domino. Will the whole line of dominoes fall over?

There are two questions we need to ask: does the pea knock the first domino over and is the distance apart of the dominoes small enough that each domino falling will knock over its neighbour?

A very useful technique of mathematical proof depends on these two ideas:

1. Can you get to the first step?
2. Can you get from any step to the next?

## ● *Example 35*

Here are some statements, with some evidence to support them:

(a) $1^2 + 2^2 + 3^2 + \ldots + n^2 = \frac{1}{6}\{n(n+1)(2n+1)\}$
We can easily check that this statement is true for $n = 1$, for $n = 2$ and for $n = 3$.

(b) $n^2 \geq 2^{n-1}$
Again, it is easy to check a few special cases, for example the statement is true for $n = 1$, for $n = 2$ and for $n = 3$.

(c) $1 + 2 + 3 + \ldots + n = \frac{1}{2}\{n(n+1)\}$
Again, this statement is easily seen to be true when $n = 1, 2$ or $3$.

Which of these statements are true for all values of $n$? How many values of $n$ do we need to check? In fact, examples (a) and (c) are true for all integers $n$ but example (b) is false for $n \geq 7$.

Consider Example 35(a). Suppose we know that for some particular value of $n$,

$$1^2 + 2^2 + 3^2 + \ldots + n^2 = \frac{1}{6}\{n(n+1)(2n+1)\} \tag{1}$$

What can we then say about $1^2 + 2^2 + 3^2 + \ldots + n^2 + (n+1)^2$?
If we add the next term, $(n+1)^2$, to both sides of equation (1) we get

$$
\begin{aligned}
1^2 + 2^2 + 3^2 + \ldots + n^2 + (n+1)^2 &= \tfrac{1}{6}\{n(n+1)(2n+1)\} + (n+1)^2 \\
&= \tfrac{1}{6}(n+1)\{n(2n+1) + 6(n+1)\} \\
&= \tfrac{1}{6}(n+1)(2n^2 + 7n + 6) \\
&= \tfrac{1}{6}(n+1)(n+2)(2n+3) \\
&= \tfrac{1}{6}(n+1)\{(n+1)+1\}\{2(n+1)+1\}
\end{aligned}
$$

Thus,

$$1^2 + 2^2 + 3^2 + \ldots + (n+1)^2 = \frac{1}{6}(n+1)\{(n+1)+1\}\{2(n+1)+1\}$$

which is the same as equation (1), but with $n$ replaced by $n+1$. So now we have shown that if the statement holds for $n$ then it will hold for $n+1$. But we know that it holds for 3 so it must also hold for 4. And as it holds for 4 it must also hold for $5, \ldots$, and so on.

## Example 36

Imagine a country which has just two different coins, one of value 3p and the other of value 7p. Can we use these coins to purchase anything costing 12p or more?

$$12p = 3p + 3p + 3p + 3p$$
$$13p = 3p + 3p + 7p$$
$$14p = 7p + 7p$$

This looks promising. We shall try to find a proof.

Suppose we can make a purchase of $k$ pence. Could we then make $k+1$ pence? There are two cases to consider.

1.  If we get $k$ pence using at least two 7p coins, then we can replace the two 7p's by five 3p's to get $k+1$ pence.
2.  If we use at least two 3p coins to get $k$ pence then we can replace two 3p coins by one 7p coin to get $k+1$ pence.

One of these two cases must apply, for the following reason. If we use at most one 3p and at most one 7p coin, the value is at most 10p. But we are considering purchases of at least 12p, so we must use at least two coins of at least one value.

Now, we have shown that:

- we can make 12p
- for any $k \geq 12$, if we can make $k$ pence, we can also make $k+1$ pence.

This gives a proof for all values of $k \geq 12$p.

These are both examples of the principle of mathematical induction.

## The principle of mathematical induction

For each positive integer $n$, let $P(n)$ be a statement that may be either true or false. Suppose that

(i)   $P(1)$ is true
(ii)  for all positive integers $n$, if $P(n)$ is true then $P(n+1)$ is also true.

Then, $P(n)$ is true for all $n \geq 1$.

Condition (i) is called the **basis for the induction** and condition (ii) is called the **inductive step**.

In our two coin values example we are not starting at $n = 1$, and the basis for the induction is $P(12)$, i.e. it is possible to get a value of 12p. Now, the inductive step shows that if it is possible to make a value of $k$p, then it is also possible to make a value of $(k+1)$p.

## Example 37

Show that, for any integer $n \geq 1$,

$$1 + 2 + 3 + 4 + \ldots + n = \frac{1}{2}n(n+1)$$

SOLUTION

We shall prove this using the principle of mathematical induction. Let $P(n)$ be the proposition

$$1 + 2 + 3 + 4 + \ldots + n = \frac{1}{2}n(n + 1)$$

(i)  $P(1)$ means $1 = \frac{1}{2}(1)(1 + 1)$. This is true, and we have established the basis for the induction.

(ii)  If $P(n)$ is true then $1 + 2 + \ldots + n = \frac{1}{2}n(n + 1)$, and by adding $(n + 1)$ to both sides we get

$$\begin{aligned}
1 + 2 + \ldots + n + n + 1 &= \tfrac{1}{2}n(n + 1) + (n + 1) \\
&= (n + 1)(\tfrac{1}{2}n + 1) \\
&= (n + 1)\{\tfrac{1}{2}(n + 2)\} \\
&= \tfrac{1}{2}(n + 1)(n + 2)
\end{aligned}$$

and hence $P(n + 1)$ is true. This establishes the inductive step.

Thus, by the principle of mathematical induction we have the required result.

If the proof of the result we want turns out to be rather complicated, we can sometimes explain it more easily by breaking it down into simpler arguments, each of which can be established by a separate, self-contained proof. This process is illustrated by the next two examples.

## Example 38

Show that $n^2 > 2n + 1$, $\forall n \geq 3$.

SOLUTION

Let $P(n)$ be the proposition that $n^2 > 2n + 1$.

(i)  $P(3)$ states that $3^2 > 2 \times 3 + 1$, i.e. that $9 > 7$, which is true. This establishes the basis for the induction.

(ii)  If $P(n)$ is true, then

$$n^2 > 2n + 1$$

If we add $2n + 1$ to both sides of this inequality, we get

$$n^2 + 2n + 1 > 2n + 1 + 2n + 1$$

Simplifying both sides, we see that this is equivalent to

$$(n + 1)^2 > 4n + 2$$

Now, since $4n > 2n + 1$, we can write a 'chain' of inequalities,

$$(n + 1)^2 > 4n + 2 > 2n + 3$$

and since $2n + 2 = 2(n + 1)$, it follows that

$$(n + 1)^2 > 2(n + 1) + 1$$

This last statement is just $P(n + 1)$, so we have established the inductive step.

Thus, by the principle of mathematical induction, $P(n)$ is true for all $n \geq 3$.

## ● *Example 39*

Show that $2^n > n^2, \forall n \geq 5$.

SOLUTION
Let $P(n)$ be the proposition that

$$2^n > n^2$$

(i)   $P(5)$ states that $2^5 > 5^2$, i.e. that $32 > 25$, which is true.
(ii)  If $P(n)$ is true then $2^n > n^2$, and by multiplying both sides by 2 we get

$$2^{n+1} = 2 \times 2^n > 2n^2 = n^2 + n^2$$

Now, using the previous example, we know that $n^2 > 2n + 1$ for $n \geq 5$, and it follows that

$$2^{n+1} > n^2 + 2n + 1 = (n+1)^2 \text{ for } n \geq 5.$$

Hence, $P(n+1)$ is true.

Thus, by the principle of mathematical induction we have the required result.

We might ask whether we really need to prove *both* the basis and the inductive step? Both parts of the induction are important, as can be seen in the following examples.

## ● *Example 40*

Show that the statement $2 + 4 + \ldots + 2n = (n+2)(n-1)$ for all $n \geq 1$ satisfies the inductive step but has no basis.

SOLUTION
Let $P(n)$ be the proposition that $2 + 4 + \ldots + 2n = (n+2)(n-1)$.

(i)   $P(1)$ means that $2 = (1+2)(1-1)$, which is not true.
(ii)  If $P(n)$ were true, then $2 + 4 + \ldots + 2n = (n+2)(n-1)$ would be true, and by adding $2(n+1)$ to both sides we would get

$$2 + 4 + \ldots + 2n + 2(n+1) = (n+2)(n-1) + 2(n+1)$$
$$= (n+3)n = ((n+1)+2)((n+1)-1)$$

Hence, $P(n+1)$ would also be true.

Thus the inductive step is satisfied, but the basis for the induction fails and the result is false.

## ● *Example 41*

Let $N = n^2 + n + 41$. Show that there are some values of $n$ for which $N$ is a prime number, and others for which it is not. It follows that there is no inductive step which would show that $n^2 + n + 41$ is a prime number for all possible $n$.

SOLUTION

Let $P(n)$ mean that $n^2 + n + 41$ is a prime number.

(i)   $P(1)$ means that $1^2 + 1 + 41$ is a prime. This is true, since 43 is prime.
(ii)  You have to work quite hard to find a value of $n$ for which $n^2 + n + 41$ is not a prime number! In fact, $P(1), P(2), \ldots, P(39)$ are all true, but $P(40)$ and $P(41)$ are false. For example, if $n = 39$, then $N = 39^2 + 39 + 41 = 1601$, which is prime. However, if $n = 40$ then $N = 40^2 + 40 + 41 = 1681$, and $1681 = 41^2$. Hence, $N$ is not prime when $n = 40$, the truth of $P(39)$ does not imply the truth of $P(40)$ and no inductive step is possible.

Thus, the basis for the induction holds but the inductive step fails and the result is false.

## EXERCISES ON 1.6

1.   Show that with a supply of 3p and 5p stamps you can make up all postages of 8p or more.
2.   Prove that $2^n \geq n^2$ for all integers $n \geq 4$.
3.   Prove each of the following statements, using the principle of mathematical induction.
     (a) $1^3 + 2^3 + 3^3 + \ldots + n^3 = \{(\frac{1}{2})n(n+1)\}^2$ for all $n \geq 1$.
     (b) $6^n - 5n + 4$ is divisible by 5 for all $n \geq 1$.
     (c) $1 + a + a^2 + a^3 + a^4 + \ldots + a^n = (a^{n+1} - 1)/(a - 1)$ for all $n \geq 1$.
     (d) $(1 + 2 + \ldots + n)^2 = 1^3 + 2^3 + \ldots + n^3$ for all $n \geq 1$.
4.   A class of students is told that there will be a surprise fire-drill one day the following week. They reason that the drill cannot be left until the Friday, because then it would no longer be a surprise. But if we accept that it cannot be on the Friday, then it cannot be on Thursday either, because again it would no longer be a surprise. Continuing the argument, they reach the conclusion that there can be no such thing as a surprise fire-drill. Express the students' argument in formal logic, and find the fallacy in the argument.

# 1.7 Project

The Fibonacci sequence is the sequence $1, 1, 2, 3, 5, 8, 13, \ldots$. Each term is the sum of the two preceding terms. Any number which appears in this sequence is called a Fibonacci number.

Leonardo of Pisa (*c*. 1170–1250), commonly known as Fibonacci (son of Bonacci), an Italian mathematician who introduced the Arabic number system to Europe, posed the following problem in his book, *Liber Abaci*, published in 1202:

'One pair of rabbits are put in a certain place surrounded by a wall. How many rabbits can be produced from that pair in a year, if the nature of these rabbits is such that every month each pair bears a new pair which from the second month on becomes productive?'

Explain why the total numbers of rabbits each month follow the Fibonacci sequence.

**Fig 1.2** Leonardo Fibonacci (David Eugene Smith Collection, Rare Book and Manuscript Library, Columbia University)

We can use the principle of mathematical induction to establish many interesting properties of the Fibonacci sequence.

Can you write each of the numbers 1 to 10 as sums of distinct Fibonacci numbers? For example:

$$6 = 5 + 1$$

Can you also express 100 as a sum of distinct Fibonacci numbers?

To express 100 as a sum of Fibonacci numbers, we find the largest Fibonacci number less than 100. This is 89. Now, $100 - 89 = 11$, so our next task is to express 11 as a sum of Fibonacci numbers. We do this by noticing that $11 = 8 + 3$. This shows that $100 = 89 + 8 + 3$, which expresses 100 as a sum of three Fibonacci numbers.

Can *every* positive integer be expressed as a sum of distinct Fibonacci numbers? Try to give a proof that all numbers can be expressed in this way.

## HINT

You are probably used to solving mathematical problems where all of the notation is defined for you. For this problem, *you* have to decide what notation to use, and it is a good idea to name things in ways which

- agree with existing notation for general concepts
- act as mnemonics for specific objects.

So, we will use $f(n)$ to denote the $n$th Fibonacci number ('f' for 'Fibonacci') and $P(n)$ to mean that all positive integers less than $f(n)$ can be written as the sum of distinct numbers chosen from $f(1), f(2), \ldots, f(n-1)$.

### Subsequences

2, 8, 34, 144, 610, 2584, 10946, 46368, 196418, 832040, . . .

This sequence is obtained by taking every third term of the Fibonacci sequence.

If we form a new sequence by taking just some of the terms of a particular sequence we call this new sequence a *subsequence* of the original sequence. For example, we can describe the above subsequence as $A(n)$, where $A(n) = f(3n)$ (in this case, the '$A$' doesn't stand for anything in particular—it's just the first letter of the alphabet).

The first few terms of this new sequence, $A(1), A(2), \ldots, A(10)$, are all even numbers. Must this be true for all $A(n)$?

We can rewrite $A(2)$ as

$$
\begin{aligned}
A(2) &= 8 \\
&= 5 + 3 \\
&= f(5) + f(4) \\
&= \{f(4) + f(3)\} + f(4) \\
&= 2f(4) + f(3) \\
&= 2 \times 3 + A(1)
\end{aligned}
$$

Similarly,

$$
\begin{aligned}
A(3) &= 34 \\
&= 21 + 13 \\
&= f(8) + f(7) \\
&= \{f(7) + f(6)\} + f(7) \\
&= 2f(7) + f(6) \\
&= 2 \times 13 + A(2)
\end{aligned}
$$

Find $A(n + 1) - A(n)$ for general $n$.

Now prove that every term of $A(n)$ will be divisible by 2.

Another interesting subsequence of the Fibonacci sequence is defined by selecting every fourth term. Call this new sequence $B(n)$. Then,

$$B(n) = f(4n)$$

Write down the first six terms of the sequence $B(n)$. Is there a common factor amongst these terms? Write down expressions for $B(2)$, for $B(3)$ and for $B(n + 1) - B(n)$. Use these results to prove that every fourth term of the Fibonacci sequence is divisible by 3, i.e. that for any value of $n$, the Fibonacci number $f(4n)$ is divisible by 3.

Is there a similar subsequence of Fibonacci numbers all of which are divisible by 5?

## *Summary*

A simple proposition is usually called a **primitive proposition**. Logical operators can be used to combine primitive propositions to obtain **compound propositions**. The logical operators are **not** for negation, **or** for disjunction, **and** for conjunction, $\Rightarrow$ for if then, $\Leftrightarrow$ for if and only if. The symbols $T$ and $F$ denote the truth values true and false respectively.

Truth tables for these logical operators are :

| $p$ | **not** $p$ |
|-----|-------------|
| $T$ | $F$ |
| $F$ | $T$ |

| $p$ | $q$ | $p$ **and** $q$ |
|-----|-----|-----------------|
| $T$ | $T$ | $T$ |
| $T$ | $F$ | $F$ |
| $F$ | $T$ | $F$ |
| $F$ | $T$ | $F$ |

| $p$ | $q$ | $p$ **or** $q$ |
|-----|-----|----------------|
| $T$ | $T$ | $T$ |
| $T$ | $F$ | $T$ |
| $F$ | $T$ | $T$ |
| $F$ | $F$ | $F$ |

| $p$ | $q$ | $p \Rightarrow q$ |
|-----|-----|-------------------|
| $T$ | $T$ | $T$ |
| $T$ | $F$ | $F$ |
| $F$ | $T$ | $T$ |
| $F$ | $F$ | $T$ |

| $p$ | $q$ | $p \Leftrightarrow q$ |
|-----|-----|-----------------------|
| $T$ | $T$ | $T$ |
| $T$ | $F$ | $F$ |
| $F$ | $T$ | $F$ |
| $F$ | $F$ | $T$ |

The following all mean the same thing: $p \Rightarrow q$; $p$ is **sufficient** for $q$; $q$ is **necessary** for $p$. Also, $p \Leftrightarrow q$ means the same as $p$ is **necessary** and **sufficient** for $q$. This is often written $p$ iff $q$.

Propositions that are always true are called **tautologies**. Propositions that are always false are called **contradictions**.

We use the symbol $\forall$ to denote the logical quantifier **for all**, and the symbol $\exists$ to denote the logical quantifier **there exists**.

We have considered six different types of proof: direct proof, contrapositive proof, proof by contradiction, biconditional, counterexample and mathematical induction.

**The principle of mathematical induction** is as follows. For each positive integer $n$, let $P(n)$ be a statement that may be either true or false. Suppose that

(i)   $P(1)$ is true,
(ii)  for all positive integers $n$, if $P(n)$ is true then $P(n + 1)$ is also true,

then $P(n)$ is true for all $n \geq 1$.

Condition (i) is called the **basis for the induction**. Condition (ii) is called the **inductive step**.

# $2 \bullet$ Sets

Set theory only evolved during the nineteenth century, but now pervades every branch of modern mathematics. Georg Cantor (1845–1918), in 1895, was the first mathematician to define a set formally. Cantor, who lived most of his life in Germany, went on to develop the branch of mathematics now known as set theory. His views encountered strong opposition from some of his contemporaries. However, in the judgement of David Hilbert (1862–1943), another important German mathematician of that period,

'No one shall expel us from the paradise which Cantor has created for us.'

**Fig 2.1** Georg Cantor (David Eugene Smith Collection, Rare Book and Manuscript Library, Columbia University)

**Fig 2.2** David Hilbert (David Eugene Smith Collection, Rare Book and Manuscript Library, Columbia University)

In this chapter, we define what is meant by a *set*, and describe the various ways in which sets can be combined to form new sets. This leads us to study rules for manipulating sets which are analogous to the rules we use to manipulate numbers and algebraic expressions. We end the chapter with a discussion of finite and infinite sets, and demonstrate that there are two different kinds of infinite sets, called *countable* and *non-countable*.

## 2.1 Introduction

A **set** is a collection of specified objects.

## ● *Example 1*

Consider five sets of objects. This example shows that the objects which make up a set may be physical objects, numbers, letters of the alphabet, words which identify a class of objects, or a mixture of any of these.

- {Pyramid of Cheops, Great Wall of China, Empire State Building}
- {1, 2, 3, 4}
- {a, e, i, o, u}
- {spaghetti, rice, bread, potato}
- {1, 2, e, Great Wall of China}

The objects which make up a set are usually called **elements**, or **members**, of the set. These elements are displayed between **braces** { }. The braces are often read as

the set whose elements are . . .

Also, we usually use capital letters to identify individual sets. In the first example above, if we want to name the set '*A*', we would write

$$A = \{1, 2, 3, 4\}$$

to be read as:

*A* is the set whose elements are 1, 2, 3 and 4.

The order in which we present the elements in a set is not important. For example, the sets

$$A = \{1, 3, 5, 7\}$$

and

$$B = \{3, 1, 5, 7\}$$

are, in fact, the same set.

We also do not allow a set to have duplicate elements. For example, if in a classroom there are 10 children aged 6, and 15 children aged 7, then the set, *C* say, of children in the class has 25 elements. But we could also define a set of *children's ages*, *A* say, and this would have only two elements, $A = \{6, 7\}$.

If a set is small we can easily write out all of its elements explicitly, as in {1, 3, 5, 7}. If a set is large, it would be tedious to list all of its elements, and there may even be an infinite number of them. However, we are usually most interested in sets whose elements are related to each other in some systematic way, and we would then list the elements implicitly. For example, the set of all prime numbers can be written as

$$\{x : x \text{ is a prime number}\}$$

to be read as

the set whose elements are all *x* such that *x* is a prime number

or, slightly less clumsily,

the set of all prime numbers.

Note that the colon is read as *such that*. In simple examples, it seems an unnecessarily elaborate notation, but it helps to avoid confusion in more complex examples.

## ● *Example 2*

The set $X$ whose elements are the even integers can be written as

$$\{\ldots, -4, -2, 0, 2, 4, \ldots\}$$

where the dots indicate that the sequence continues indefinitely in each direction. In this example, it is reasonably clear that this is what the sequences of dots mean, but a more precise way to write $X$ would be as

$$X = \{x : x \text{ is an even integer}\}$$

What might we mean by the following statement?

$$Y = \{2, 4, 8, 16, \ldots\}$$

One possible interpretation is

$$Y = \{x : x = 2^n, \ n \text{ a natural number}\}$$

Another (admittedly of rather less obvious interest!) is

$$Y = \{x : x = (n^3 - 3n^2 + 8n)/3, \ n \text{ a natural number}\}$$

Check for yourself that these two definitions of $Y$ agree if $n = 1, 2, 3, 4$ but disagree when $n = 5$.

Suppose we define two sets $A$ and $B$ as

$$A = \{2, 4, 6, 8\}$$

and

$$B = \{x : x \text{ is a prime number}\}$$

Then, 4 is an element of $A$. We write this as $4 \in A$, using the symbol $\in$ to represent 'is an element of'. Also, 4 is **not** an element of $B$. We write this as $4 \notin B$ where $\notin$ is read as 'is not an element of'.

When we are classifying objects, we need to decide how detailed to make the classification. For example, if we define a set $A$ as

$$A = \{x : x \text{ is a student}\}$$

then, in principle, we could classify any person as either an element of $A$ or not an element of $A$. Similarly, if we define

$$B = \{x : x \text{ is a student of mathematics}\}$$

then again, in principle any person is either an element of $B$ or not an element of $B$. But we can also say the following:

if $x \in B$, then it follows automatically that $x \in A$

This is because $B$ is, in a reasonable sense, contained within $A$. We say that $B$ is a **subset** of $A$. Another example would be that $B = \{1, 5\}$ is a **subset** of $A = \{1, 3, 5\}$.

We write this as $B \subseteq A$. We allow a subset to be equal to the original set. So, if $A$ is any set, then it is always the case that $A \subseteq A$. In general, we say that:

the set $A$ is a **subset** of the set $B$ if every element of $A$ is also an element of $B$.

More formally,

$$A \subseteq B \Leftrightarrow (\forall x, x \in A \Rightarrow x \in B)$$

If we want to talk about a subset of $B$ which does not include the whole set then we talk about a **proper subset** of $B$.

## Example 3

Two examples of subsets:

$$\{1, 5\} \subseteq \{1, 2, 3, 4, 5\}$$
$$\{1, 5\} \subseteq \{1, 5\}$$

Strictly, we need to distinguish between an element of a set, and a set with one element.

## Example 4

Let $A = \{1, 2\}$, then $1 \in A$ but $\{1\} \subseteq A$.

We shall say that two sets $X$ and $Y$ are **equal** if they have the same elements.

## Example 5

Which of the following sets are equal?

- $A = \{-1, 1, 2\}$
- $B = \{-1, 2, 1\}$
- $C = \{0, 1, 2\}$
- $D = \{2, 1, -1, -2\}$
- $E = \{x : x^2 = 4 \text{ or } x^2 = 1\}$

SOLUTION
Sets $A$ and $B$ are equal, because the order of the elements does not matter. Sets $D$ and $E$ are equal—although they are described differently, they contain the same elements. Set $C$ does not equal $A$ or $B$ or $D$ or $E$.

## Example 6

Are the following two sets equal?

- $X = \{x : x^2 - 5x + 6 = 0\}$
- $Y = \{2, 3\}$

SOLUTION
$x^2 - 5x + 6 = 0$ has two solutions, $x = 2$ and $x = 3$. The elements of $X$ are therefore the same as the elements of $Y$, and $X = Y$.

A more formal way to express equality of two sets is:

$$A = B \Leftrightarrow (x \in A \Rightarrow x \in B) \text{ and } (x \in B \Rightarrow x \in A)$$

In set notation, this is equivalent to $A \subseteq B$ and $B \subseteq A$. If we need to prove that two sets, $A$ and $B$ say, are equal, and the result is not obvious by inspection, we have to prove two things – that $A \subseteq B$, and that $B \subseteq A$.

In principle, we can count the number of elements in any set. We write $|X|$ for the number of elements in $X$. In Example 6, $|X| = 2$, whereas in Example 2, the set $X$ has an infinite number of elements. We write this as $|X| = \infty$.

We shall often need the following sets:

$\mathbb{Z} = \{\text{integers}\} = \{\ldots, -2, -1, 0, 1, 2, \ldots\}$ ( the letter $\mathbb{Z}$ is used because it stands for *Zahlen*, the German word for integers),

$\mathbb{N} = \{\text{natural numbers}\} = \{1, 2, \ldots\}$.

## EXERCISES ON 2.1

1. Express the following sets explicitly:
   (a) $\{x : x \in \mathbb{Z} \text{ and } -2 < x < 5\}$
   (b) $\{x : x \text{ is a day of the week with no letter } n\}$
   (c) $\{x : x^2 = 1\}$
   (d) $\{x : x \in \mathbb{Z} \text{ and } x^4 = 1\}$

2. Write the following sets using set notation:
   (a) $\{0, 2, 4, 6, 8, \ldots\}$
   (b) $\{\ldots, -2, -1, 0, 1, 2, \ldots\}$
   (c) $\{0, 1, 2, 3\}$
   (d) $\{0, 3, 6, 9, \ldots\}$

3. Which of the following pairs of sets are equal?
   (a) $\{1, 2, 3\}, \{2, 1, 3\}$
   (b) $\{x : x^2 + 2x + 1 = 0\}, \{1, -1\}$
   (c) $\{2, 3, 5\}, \{x : x \text{ is a prime number less than } 8\}$

4. Give a simpler description of the sets
   (a) $\{x : x \text{ is a positive even number and } x^2 + 4x = 12\}$
   (b) $\{x : x \text{ is an even prime number greater than } 3\}$

5. True or false?
   (a) $2 \in \{2, 3, 4\}$
   (b) $2 \in \{2\}$
   (c) $\{2\} \in \{2, 3, 4\}$
   (d) $\{2, 3\} \in \{1, 3, \{2, 3\}\}$
   (e) $\{1, 2\} \in \{1, 2, 3, 4\}$
   (f) $1 \in \{1, 2\}$
   (g) $\{1, 2\} \subseteq \{1, \{1, 2\}\}$

6. True or false? For those statements that are false give a counterexample.
   (a) If $A \subseteq B$ and $B \subseteq C$ then $A \subseteq C$.
   (b) If $A \subseteq B$ then $B \subseteq A$.
   (c) If $A \in B$ and $B \subseteq C$ then $A \in C$.
   (d) If $A = B$ then $A \subseteq B$.
   (e) If $A \notin B$ and $B \notin C$ then $A \notin C$.

7.  What is the connection between the elements of this set? Can you find another element of the set?

    {pain, ran, ales, man, inland}

8.  For each of the following sets $X$, evaluate $|X|$:

    (a) $X = \{$Monday, Tuesday, Wednesday, Thursday, Friday$\}$

    (b) $X = \{x : x \in \mathbb{Z}$ and $x^2 = x\}$

    (c) $X = \{x : x \in \mathbb{N}$ and $x^2 \leq 25x\}$

    (d) $X = \{x : x \in \mathbb{Z}$ and $x$ is a prime number$\}$

# 2.2 Operations on sets

Given any two sets, there are various different operations we can use to combine them into new sets. We shall use the four sets in the following example to illustrate these new operations.

## ● Example 7

Consider four sets:

- $R = \{x : x$ is an odd, positive integer less than 6$\}$
- $S = \{x : x^2 - 7x + 10 = 0\}$
- $T = \{x : x$ is an even positive integer less than 6$\}$
- $W = \{2, 4, 5\}$.

You may wish to list explicitly the elements of $R, S$ and $T$.

We can combine two sets $A$ and $B$ to get a new set, called the **union of $A$ and $B$**, $A \cup B$, which we define as

$$A \cup B = \{x : x \in A \text{ or } x \in B\}$$

## ● Example 8

Using the sets $R, S$ and $T$ from Example 7 we have:

$$R \cup T = \{1, 2, 3, 4, 5\}$$
$$R \cup S = \{1, 2, 3, 5\}$$

Notice that an element $x$ is included in $A \cup B$ if it is an element of $A$, or an element of $B$, or an element of both $A$ and $B$. Similarly, we can look at the overlap of two sets $A$ and $B$, called the **intersection of $A$ and $B$**, $A \cap B$, which we define as

$$A \cap B = \{x : x \in A \text{ and } x \in B\}$$

## ● Example 9

Using the sets $R, S$ and $T$ from Example 7 we have:

$$R \cap S = \{5\}$$
$$S \cap T = \{2\}$$

We can also list the elements of $A$ which are not in $B$. This operation is called the **difference between $A$ and $B$**, which we write as $A \backslash B$. Thus,

$$A \backslash B = \{x : x \in A \text{ and } x \notin B\}$$

## Example 10

Using the sets $R$, $S$ and $T$ from Example 7 we have

$$R \backslash T = \{1, 3, 5\}$$
$$R \backslash S = \{1, 3\}$$

Because the result of any of these operations is another set, we can combine as many operations as we like:

## Example 11

Using the sets $R$, $S$, $T$ and $W$ from Example 7 we have

$$(R \cup S) \cup W = \{1, 2, 3, 5\} \cup \{2, 4, 5\} = \{1, 2, 3, 4, 5\}$$
$$(R \cap W) \cup S = \{5\} \cup \{2, 5\} = \{2, 5\}$$

### The empty set and the universal set

A set with nothing in it is called the **empty set**. We write this as $\{\}$ or $\emptyset$ where $\emptyset$ is a special symbol used to denote the empty set. On the face of it, a set with nothing in it is not very interesting. We define it mainly so that when we define more complicated ideas to do with sets, we do not have to qualify everything we say by statements like 'provided the resulting set has something in it'.

## Example 12

Using the sets from Example 7, explain why $R \cap T = \emptyset$.

SOLUTION
$R \cap T = \emptyset$ because no positive integer can be both odd and even.

We also use a special name for the 'biggest' set we are prepared to consider. If, in a particular problem, all of the sets that we want to consider are subsets of a big set $\mathcal{U}$, then $\mathcal{U}$ is called the **universal set**. This definition depends on the application (unlike the empty set, which always means the same thing).

For example, our university department teaches ten first-year courses in mathematics, and students can choose to do all ten, or one of several selections of five from ten. So if we want to discuss the possible sets of *mathematics courses* taken by our students, all sets are subsets of a universal set $\mathcal{U}$ with 10 elements, corresponding to the ten courses on offer. But the university also allows students to combine mathematics courses with courses in other departments. So if we wanted to discuss the sets of *courses* taken by our students, then $\mathcal{U}$ would be a much bigger set.

## • *Example 13*

Using the sets $R$ and $T$ from Example 7, if

$\quad \mathcal{U} = \{x : x \text{ is a positive integer less than } 6\}$

then $R \cup T = \mathcal{U}$.

The set of elements not in $A$ is called the **complement of $A$**, and is written as $A^c$. Thus,

$\quad A^c = \{x : x \notin A\}$

Note that $A^c$ is only well-defined if we specify the universal set $\mathcal{U}$.

## • *Example 14*

Let $\mathcal{U} = \{1, 2, 3, 4, 5\}$ and consider the two sets $A = \{1, 2\}$ and $B = \{3\}$. Then

$\quad A^c = \{3, 4, 5\}$ and $B^c = \{1, 2, 4, 5\}$

## • *Example 15*

Let $\mathcal{U} = \mathbb{N}$, and consider the two sets $A = \{1, 2\}$ and $B = \{3\}$ as before. Then

$\quad A^c = \{3, 4, 5, 6, \ldots\}$ and $B^c = \{1, 2, 4, 5, 6, \ldots\}$

## • *Example 16*

We can see that the following results are true:

- $\emptyset^c = \mathcal{U}$
- $\mathcal{U}^c = \emptyset$
- $(A^c)^c = A$
- If $A \subseteq B$ then $B^c \subseteq A^c$.

## • *Example 17*

What are the sets $A \cup A, A \cup \emptyset, A \cap A, A \cap \emptyset$?

SOLUTION

$\quad A \cup A = A, A \cup \emptyset = A, A \cap A = A, A \cap \emptyset = \emptyset.$

All of these properties can be illustrated easily using diagrams known as **Venn diagrams**. The following example shows how to use a Venn diagram.

## • *Example 18*

Suppose we are considering a set of students, and we want to identify the sets

- $A = \{\text{students who wear glasses}\}$
- $B = \{\text{students who enjoy mathematics}\}$

Then, we can represent the universal set of all students as a rectangle, and the sets $A$ and $B$ as circles. A student who wears glasses and enjoys mathematics will be in $A \cap B$, which is represented in Fig 2.3 by the region where the circles overlap.

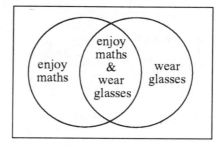

**Fig 2.3**

Venn diagrams were named after John Venn (1834–1923), an English mathematician. Figure 2.4 shows a stained glass window commemorating John Venn in Gonville and Caius College, Cambridge. The window shows a Venn diagram for three overlapping sets.

**Fig 2.4** John Venn stained glass window in the Hall of Gonville and Caius College, Cambridge

Since Venn's first diagrams, mathematicians have been searching for good representations of four or more intersecting sets. Anthony Edwards, a fellow of Gonville and Caius College, Cambridge, has recently produced symmetrical drawings for arbitrary numbers of sets. For more information on his diagrams see his article 'Venn diagrams for many sets' in *New Scientist*, 7 January 1989, pages 51–56. Figure 2.5 shows the Edwards diagram for five sets.

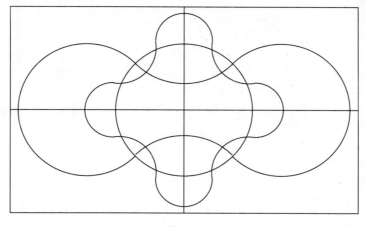

**Fig 2.5**

The hatched parts of the Venn diagrams in Fig 2.6 illustrate the sets $A \cup B$, $A \cap B$, $A \backslash B$ and $A^c$.

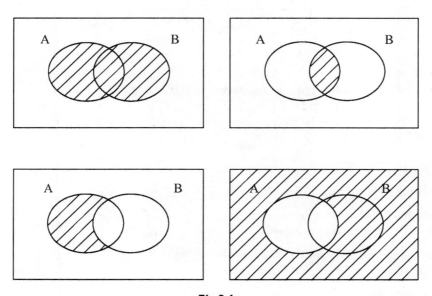

**Fig 2.6**

## ● *Example 19*

Given the information that $A \cup B = \{1, 2, 3, 4\}$, $A \cap B = \{3\}$, $A \backslash B = \{1, 2\}$, $A^c = \{4, 5, 6\}$, we can put the numbers $\{1, 2, 3, 4, 5, 6\}$ in their correct places in the Venn diagram of Fig 2.7.

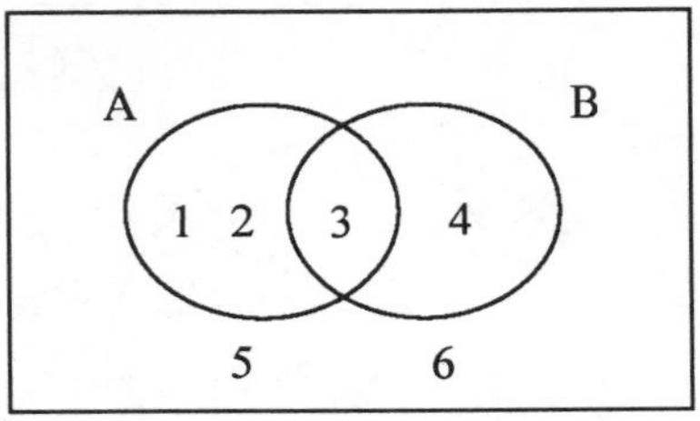

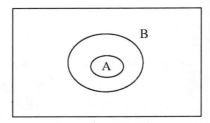

**Fig 2.7**

## Example 20

If $A \subseteq B$ what are $A \cup B$ and $A \cap B$?

**Fig 2.8**

SOLUTION
From Fig 2.8 we can see that if $A \subseteq B$ then $A \cup B = B$ and $A \cap B = A$.

Operations on sets can be defined in many different ways, but some of these give equivalent answers.

## Example 21

Let $\mathcal{U}$ be the set of all positive integers $x \leq 10$, and define

- $A = \{x : x \text{ is a prime number}\}$,
- $B = \{x : x \text{ is an odd number}\}$.

Find the sets $(A \cup B)^c$ and $A^c \cap B^c$.

SOLUTION
First, note that $x \in A \cup B$ if $x$ is either prime or odd, hence

$$A \cup B = \{1, 2, 3, 5, 7, 9\}$$

Then, $(A \cup B)^c$ consists of all elements of $\mathcal{U}$ which are not in $A \cup B$, hence

$$(A \cup B)^c = \{4, 6, 8, 10\}$$

Similarly,

$$A^c = \{1, 4, 6, 8, 9, 10\}$$

and

$$B^c = \{2, 4, 6, 8, 10\}$$

Now, $A^c \cap B^c$ consists of the integers $x \leq 10$ which are in both $A^c$ and $B^c$. Hence,

$$A^c \cap B^c = \{4, 6, 8, 10\}$$

Notice that in the last example, $(A \cup B)^c = A^c \cap B^c$. Try the same example with $\mathcal{U}$ as before, but now take $A = \{1, 4, 7, 10\}$ and $B = \{1, 3, 5, 6, 7\}$. What do you see? Might this be a general result?

Remember that if we want to prove that two sets $X$ and $Y$ are equal, we need to show two things:

1. $X \subseteq Y$
2. $Y \subseteq X$

We shall illustrate this by the following example.

## Example 22

Show that if $A \subseteq B$, then $A \cup B = B$.

SOLUTION
We first show that $A \cup B \subseteq B$:

Let $x \in A \cup B$. Then $x \in A$ or $x \in B$ (*definition of* $\cup$).
If $x \in B$ then we are done.
If $x \in A$ then $x \in B$ (*since* $A \subseteq B$).
Thus, in either case $x \in B$.

We now show that $B \subseteq A \cup B$:

Let $x \in B$. Then $x \in A$ or $x \in B$ (*definition of 'or'*).
Thus, $x \in A \cup B$ (*definition of* $\cup$).

Therefore, the statements $A \cup B \subseteq B$ and $B \subseteq A \cup B$ are both true, and we have proved that $A \cup B = B$.

## Associativity

When we add the three numbers 3, 4 and 5 together we know that the order in which we perform the additions does not matter. If we want to, we can show the different ordering of the additions by using brackets. Thus, $(3 + 4) + 5$ means:

first add 3 and 4 to get 7, then add 5, to get 12. Thus, $(3 + 4) + 5 = 7 + 5 = 12$.

On the other hand, $3 + (4 + 5)$ means:

add 3 to the sum of 4 and 5. Thus, $3 + (4 + 5) = 3 + 9 = 12$.

And of course, as we already knew, we can see that

$$(3 + 4) + 5 = 3 + (4 + 5)$$

However, for division this property does not hold. For example,

$$(3 \div 4) \div 5 = \frac{3}{4} \div 5 = \frac{3}{20}$$

whereas

$$3 \div (4 \div 5) = 3 \div \frac{4}{5} = 3 \times \frac{5}{4} = \frac{15}{4}$$

So clearly,

$$(3 \div 4) \div 5 \neq 3 \div (4 \div 5)$$

We say that addition is **associative**, since

$$a + (b + c) = (a + b) + c \text{ for all } a, b, c$$

Similarly, division is **not** associative, because this property does not hold.

## Associative law for union and intersection

The union operation is associative, i.e. for any sets $A$, $B$ and $C$,

$$(A \cup B) \cup C = A \cup (B \cup C)$$

We can draw a Venn diagram to convince ourselves that this is true, but *remember that a picture is not a proof*.

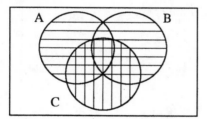 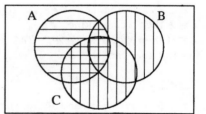

**Fig 2.9**

In the left-hand part of Fig 2.9, we have hatched $A \cup B$ horizontally and $C$ vertically. Then, the total hatched area represents the union of these two sets, i.e. $(A \cup B) \cup C$. In the right-hand diagram, we have hatched $A$ horizontally and $B \cup C$ vertically. Again, the total hatched area represents the union of the two sets, i.e. $A \cup (B \cup C)$. Although the patterns of hatching are different in the two diagrams, we can see that in each case the total hatched area is the same.

## ● Theorem I

$$(A \cup B) \cup C = A \cup (B \cup C).$$

PROOF
To show that $(A \cup B) \cup C \subseteq A \cup (B \cup C)$, we start by taking any $x \in (A \cup B) \cup C$. Then, either $x \in A \cup B$ or $x \in C$. But $x \in A \cup B$ means that either $x \in A$ or $x \in B$.

Hence, $x \in A$ or $x \in B$ or $x \in C$. That is, $x \in A$ or $x \in B \cup C$, which means that $x \in A \cup (B \cup C)$. We have now proved that

$$(A \cup B) \cup C \subseteq A \cup (B \cup C) \tag{1}$$

To show that $A \cup (B \cup C) \subseteq (A \cup B) \cup C$, we start by taking any $x \in A \cup (B \cup C)$. Then, either $x \in A$ or $x \in (B \cup C)$. But $x \in (B \cup C)$ means that $x \in B$ or $x \in C$. So $x \in A$ or $x \in B$ or $x \in C$. That is, $x \in A \cup B$ or $x \in C$, which means that $x \in (A \cup B) \cup C$. We have now proved that

$$A \cup (B \cup C) \subseteq (A \cup B) \cup C \tag{2}$$

The combination of (1) and (2) proves that $A \cup (B \cup C) = (A \cup B) \cup C$.

The intersection operation is also associative. The proof of this is left as an exercise.

## ● *Theorem 2*

For any sets $A$, $B$ and $C$, $(A \cap B) \cap C = A \cap (B \cap C)$.

## Distributivity

We are used to the following arithmetic result:

$$2 \times (3 + 4) = (2 \times 3) + (2 \times 4)$$

We say that **multiplication is distributive over addition.** In general,

$$a \times (b + c) = (a \times b) + (a \times c) \text{ for all } a, b, c$$

In contrast, addition is *not* distributive over multiplication. For example,

$$2 + (3 \times 4) \neq (2 + 3) \times (2 + 4)$$

## Distributive law for intersection over union and for union over intersection

## ● *Example 23*

Let $A$, $B$ and $C$ be the sets $A = \{1, 3\}$, $B = \{1, 4\}$ and $C = \{2, 4\}$. Then

$$A \cap (B \cup C) = A \cap \{1, 2, 4\} = \{1\}$$

and

$$(A \cap B) \cup (A \cap C) = \{1\} \cup \emptyset = \{1\}$$

In general, if we draw a Venn diagram of three sets, $A$, $B$ and $C$, we can easily convince ourselves that $A \cap (B \cup C) = (A \cap B) \cup (A \cap C)$. In the left-hand part of Fig 2.10, we have hatched $A$ horizontally and $B \cup C$ vertically. The intersection of the two sets, $A \cap (B \cup C)$, is then represented by the cross-hatched area. In the right-hand diagram, we have hatched $A \cap B$ horizontally and $A \cap C$ vertically. Now the union, $(A \cap B) \cup (A \cap C)$, is represented by the total hatched area, and we can see that this is the same as the cross-hatched area in the left-hand diagram. To prove this result, we proceed as follows.

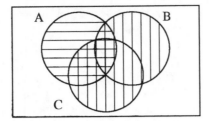 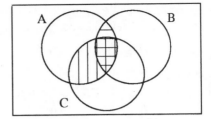

**Fig 2.10**

● *Theorem 3* ─────────────────────────────

$A \cap (B \cup C) = (A \cap B) \cup (A \cap C)$.

PROOF

We first prove that $A \cap (B \cup C) \subseteq (A \cap B) \cup (A \cap C)$. Let $x \in A \cap (B \cup C)$. Then, $x$ is in $A$ and $x$ is in $B \cup C$. So $x$ is in $A$, and $x$ is either in $B$ or in $C$. Hence, either $x$ is in $A$ and $x$ is in $B$, or $x$ is in $A$ and $x$ is in $C$. Thus, $x \in (A \cap B)$ or $x \in (A \cap C)$. Hence, $x \in (A \cap B) \cup (A \cap C)$. This proves that

$$A \cap (B \cup C) \subseteq (A \cap B) \cup (A \cap C) \tag{1}$$

We now prove that $(A \cap B) \cup (A \cap C) \subseteq A \cap (B \cup C)$. Let $x \in (A \cap B) \cup (A \cap C)$. Then, $x$ is in $A$ and $B$ or $x$ is in $A$ and $C$. So $x$ is in $A$ and $x$ is in either $B$ or $C$. Hence, $x \in A \cap (B \cup C)$. This proves that

$$(A \cap B) \cup (A \cap C) \subseteq A \cap (B \cup C) \tag{2}$$

The combination of (1) and (2) proves that $A \cap (B \cup C) = (A \cap B) \cup (A \cap C)$.

You should now be able to prove the following theorem, using a similar argument.

● *Theorem 4* ─────────────────────────────

$A \cup (B \cap C) = (A \cup B) \cap (A \cup C)$

Although we have emphasised that a Venn diagram is not a proof, the diagram is an excellent way to help you understand a problem before you try to construct a formal proof. The standard form of a Venn diagram admits the possibility that each sub-region on the diagram can have something in it. However, in some problems we know in advance that some of the sub-regions must be empty, and it is very helpful to draw our Venn diagrams in ways which convey this information.

### ● *Example 24*

This example considers Venn diagrams for mutually exclusive sets.

If $\mathcal{U}$ is the set of all university students, $A$ is the set of first-year students, and $B$ the set of second-year students, then we know that $A \cap B = \emptyset$, and the Venn diagram can be drawn as in Fig 2.11 (the standard picture is not 'wrong', it is just misleading, and therefore unhelpful).

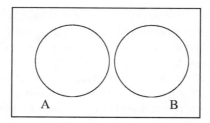

**Fig 2.11**

A slightly different situation is when $A$ is the set of all women students, and $B$ the set of all men students. Now, $A \cap B = \emptyset$ as before, but also $A \cup B = \mathcal{U}$. An appropriate Venn diagram is shown in Fig 2.12.

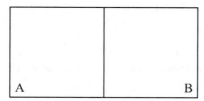

**Fig 2.12**

## EXERCISES ON 2.2

**1.** Use a Venn diagram to illustrate the identity

$$A \cup (B \cap C) = (A \cup B) \cap (A \cup C)$$

**2.** Give a counterexample which proves that, in general,

$$A \cap (B \cup C) \neq (A \cap B) \cup C$$

**3.** Use Venn diagrams to verify each of the following facts:
(a) $A \cap (B \cap C) = (A \cap B) \cap C$
(b) $A \cup (B \cap C) = (A \cup B) \cap (A \cup C)$
(c) $(A \cap B)^c = A^c \cup B^c$
(d) $A \cup A^c = \mathcal{U}$
(e) $A \cap A^c = \emptyset$
(f) $(A^c)^c = A$

**4.** Give $A^c$ when $\mathcal{U}$ is $\{x : x$ is an integer$\}$ and
(a) $A = \{x : x$ is an integer between 0 and 3 inclusive$\}$
(b) $A = \{x : x$ is an integer $\leq 10\}$
(c) $A = A_1 \cup A_2$ where $A_1 = \{x : x$ is an integer between 1 and 5 exclusive$\}$ and $A_2 = \{x : x$ is an integer between 5 and 10 exclusive$\}$

**5.** Draw a Venn diagram to represent four sets.

**6.** Give proofs for each of the following:
(a) $A \cup (B \cap C) = (A \cup B) \cap (A \cup C)$
(b) $(A \cap B) \cap C = A \cap (B \cap C)$

(It is a good idea to use Venn diagrams to convince yourself that each result is plausible, *before* you try to write out a formal proof.)

7.   What can be deduced about $A$ and $B$ if $A \cap B = A \cup B$?
8.   Draw Venn diagrams appropriate to each of the following problems.
     (a) $\mathcal{U}$ is the set of all university students, $A$ is the set of all women, $B$ is the set of all men, $C$ is the set of all mathematics students.
     (b) $\mathcal{U}$ is the set of positive rational numbers, $A$ is the set of integers, $B$ is the set of even integers, $C$ is the set of odd integers.
     (c) $\mathcal{U}$ is the set of positive rational numbers, $A$ is the set of integers, $B$ is the set of even integers, $C$ is the set of prime numbers.

## 2.3  De Morgan's laws

Augustus De Morgan was an English mathematician who died in 1871. He propounded the following conundrum:

I was $x$ years old in the year $x^2$.

Can you discover when he was born? De Morgan was also the first mathematician to work on the **four-colour problem**. The statement of this famous problem is as follows:

Is it possible to colour any map with at most four colours so that adjacent countries have different colours?

This was finally proved in the 1970s. More information on colouring maps can be found in Chapter 6.

**Fig 2.13**   Augustus De Morgan (from *Memoir of Augustus de Morgan*, published by Longmans, Green & Co.)

De Morgan's laws are

1.   The complement of the union of two sets is equivalent to the intersection of the complements of the two sets.

2.  The complement of the intersection of two sets is equivalent to the uni⊂
    the complements of the two sets.

Using set-theoretic notation, De Morgan's laws can be stated as

(i)  $(A \cup B)^c = A^c \cap B^c$
(ii) $(A \cap B)^c = A^c \cup B^c$

Before we give the proof we will illustrate (i) with an example.

## ● Example 25

Let $A = \{1,2,3\}$, $B = \{3,4,5\}$ and $\mathcal{U} = \{1,2,3,4,5,6,7\}$. Then $(A \cup B) = \{1,2,3,4,5\}$, $(A \cup B)^c = \{6,7\}$, and $A^c \cap B^c = \{4,5,6,7\} \cap \{1,2,6,7\} = \{6,7\}$. So we can see that in this example, $(A \cup B)^c = A^c \cap B^c$.

## ● Theorem 5 ───────────────────────

De Morgan's Laws. For any two sets $A$ and $B$

(i)  $(A \cup B)^c = A^c \cap B^c$
(ii) $(A \cap B)^c = A^c \cup B^c$

PROOF
To prove (i) $(A \cup B)^c = A^c \cap B^c$, we first show that $(A \cup B)^c \subseteq A^c \cap B^c$. Let $x \in (A \cup B)^c$. Then, $x$ is not in either $A$ or $B$, so $x \notin A$ and $x \notin B$. Hence $x \in A^c$ and $x \in B^c$. Thus, $x \in A^c \cap B^c$. This proves that $(A \cup B)^c \subseteq A^c \cap B^c$.

We next show that $A^c \cap B^c \subseteq (A \cup B)^c$. Let $x \in A^c \cap B^c$. Then, $x$ is not in $A$ and not in $B$, so $x$ is neither in $A$ nor in $B$. Thus $x \notin (A \cup B)$ which means that $x \in (A \cup B)^c$. This proves that $A^c \cap B^c \subseteq (A \cup B)^c$. Thus $(A \cup B)^c = A^c \cap B^c$.

The proof of (ii) follows the same lines, and is left as an exercise.

In fact, De Morgan showed that any set-theoretic identity involving only $\cap$ and $\cup$ will also be true if we swap the symbols $\cap$ and $\cup$ throughout.

## ● Example 26

Theorem 4 tells us that for any sets $A$, $B$ and $C$,

$$A \cup (B \cap C) = (A \cup B) \cap (A \cup C)$$

We can now use De Morgan's laws to prove that

$$A \cap (B \cup C) = (A \cap B) \cup (A \cap C)$$

We proceed as follows. First, using the fact that if $X = Y$ then $X^c = Y^c$ we can take complements of both sides, to give

$$(A \cup (B \cap C))^c = ((A \cup B) \cap (A \cup C))^c$$

Now, if we define $X = B \cap C$, the left-hand side of the above identity becomes

$$(A \cup X)^c$$

Using De Morgan's law (i), we have that

$$(A \cup X)^c = A^c \cap X^c = A^c \cap (B \cap C)^c$$

and using De Morgan's law (ii), we have that

$$(B \cap C)^c = B^c \cup C^c$$

So the left-hand side of the identity has become

$$A^c \cap (B^c \cup C^c)$$

Similarly, if we define $X = A \cup B$ and $Y = A \cup C$, we can use De Morgan's law (ii) on the right-hand side of the identity, to give

$$((A \cup B) \cap (A \cup C))^c = (A \cup B)^c \cup (A \cup C)^c$$

and use De Morgan's law (i) on each of $(A \cup B)^c$ and $(A \cup C)^c$ to convert the right-hand side of the identity, to give

$$(A^c \cap B^c) \cup (A^c \cap C^c)$$

Equating the new versions of the left- and right-hand sides, we obtain a new identity,

$$A^c \cap (B^c \cup C^c) = (A^c \cap B^c) \cup (A^c \cap C^c)$$

Since the names $A$, $B$ and $C$ are just labels for any sets, the same applies to $A^c$, $B^c$ and $C^c$. So we can replace $A^c$, $B^c$ and $C^c$ by $D, E$ and $F$ to obtain

$$D \cap (E \cup F) = (D \cap E) \cup (E \cap F)$$

as required.

## EXERCISES ON 2.3

1. Prove that $(A \cap B)^c = A^c \cup B^c$.
2. Use De Morgan's laws to prove that

$$A \cup (B \cup C) = (A \cup B) \cup C \Rightarrow A \cap (B \cap C) = (A \cap B) \cap C$$

3. Prove that $A \cup (B \cap A^c)^c = A \cup B^c$.
4. Prove that $((A \cap B)^c \cup B)^c = \emptyset$.

# 2.4 Power sets

Provided that a set is small, we can list all possible subsets of it.

### Example 27

Let $A = \{1, 2, 3\}$. The subsets of $A$ are $\emptyset$, $\{1\}$, $\{2\}$, $\{3\}$, $\{1, 2\}$, $\{1, 3\}$, $\{2, 3\}$ and $\{1, 2, 3\}$. Notice that $|A| = 3$ and there are $8 = 2^3$ subsets.

The set of all subsets of a set $X$ is called the **power set** of $X$, denoted by $\mathcal{P}(X)$.

## Example 28

Let $B = \{1, 2\}$. Then

$$\mathcal{P}(B) = \{\emptyset, \{1\}, \{2\}, \{1, 2\}\}$$

## Example 29

Let $C = \{1\}$. Then

$$\mathcal{P}(C) = \{\emptyset, \{1\}\}$$

## Example 30

Let $D = \{1, 3, \{1, 2, 3\}\}$. Find $\mathcal{P}(D)$.

SOLUTION

It helps if we write $C = \{1, 2, 3\}$. Then, $D = \{1, 3, C\}$, which makes it clear that $D$ is a set with three elements (although one of these elements is itself a set of three elements!). The subsets of $D$ are $\emptyset, \{1\}, \{3\}, \{C\}, \{1, C\}, \{1, 3\}, \{C, 3\}, \{1, 3, C\}$, and it follows that

$$\mathcal{P}(D) = \{\emptyset, \{1\}, \{3\}, \{\{1, 2, 3\}\},$$
$$\{1, \{1, 2, 3\}\}, \{1, 3\}, \{\{1, 2, 3\}, 3\}, \{1, 3, \{1, 2, 3\}\}\}$$

Notice again that $|D| = 3$ and there are $8 = 2^3$ subsets, hence 8 elements of $\mathcal{P}(D)$.

We shall prove that for any set $X$, $\mathcal{P}(X)$ has $2^{|X|}$ elements. As is often the case, we can see more easily how to do this if we first prove the result in a special case, say when $|X| = 3$.

## ● Theorem 6 —————————————————————

A set $X$ with 3 elements has $2^3$ subsets.

PROOF

Let the elements of $X$ be $x_1, x_2, x_3$. For any subset, $S$ say, either $x_1 \in S$ or $x_1 \notin S$, either $x_2 \in S$ or $x_2 \notin S$, and either $x_3 \in S$ or $x_3 \notin S$. So there are two possibilities for $x_1$, two for $x_2$ and two for $x_3$. So there are $2 \times 2 \times 2 = 8$ different subsets altogether.

Now we can see that the restriction to $|X| = 3$ was not essential to the argument. In general, we have the following result.

## ● Theorem 7—————————————————————

A set $X$ with $n$ elements has $2^n$ subsets.

### EXERCISES ON 2.4

1. List the power sets of the following sets:
   (a) $A = \{a, b\}$
   (b) $B = \{\{\}, 1, 2\}$
   (c) $C = \{1, 2, 3, 4\}$

2. How many elements are in $\mathcal{P}(A)$ if $A$ has
   (a) 4 elements?
   (b) 6 elements?
   (c) 10 elements?
3. Prove Theorem 7.
4. True or false?
   (a) $A \cup \mathcal{P}(A) = \mathcal{P}(A)$
   (b) $A \cap \mathcal{P}(A) = A$
   (c) $\mathcal{P}(A \cap B) = \mathcal{P}(A) \cap \mathcal{P}(B)$
   (d) $\mathcal{P}(A \cup B) = \mathcal{P}(A) \cup \mathcal{P}(B)$
5. Let $A = \{a, b\}$. List the elements of $\mathcal{P}(\mathcal{P}(B))$.
6. If $A$ has $n$ elements, $\mathcal{P}_1$ denotes the power set of $A$, $\mathcal{P}_2$ denotes the power set of $\mathcal{P}_1$, $\mathcal{P}_3$ denotes the power set of $\mathcal{P}_2$, and so on, so that $\mathcal{P}_{k+1}$ denotes the power set of $\mathcal{P}_k$ for any positive integer $k$, how many elements does $\mathcal{P}_k$ have?

## 2.5 Inclusion–exclusion

### ● Example 31

Out of 200 students, 50 of them take Mathematics, 140 take Economics, and 24 of them take both. Both courses have an exam tomorrow. Only students not taking an exam will go to the party tonight. How many students will be at the party?

SOLUTION
Let $E$ and $M$ denote the sets of Economics and Mathematics students, respectively. From the Venn diagram (Fig 2.14) we can see that if we simply add the number of students taking Economics to the number taking Mathematics, the students who are taking both will be counted twice. Thus,

$$|E \cup M| = |E| + |M| - |E \cap M|$$

So $|E \cup M| = 50 + 140 - 24 = 166$. The remaining $200 - 166 = 34$ students are taking neither Economics nor Mathematics. Thus, there will be 34 students at the party.

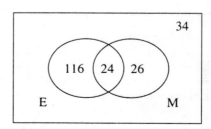

**Fig 2.14**

# Example 32

We now want to determine how many men will be at the party described in Example 31. Suppose that 60 of the 200 students are women, of whom 20 take Mathematics, 45 take Economics and 16 take both.

SOLUTION

Let $W$ denote the set of women students. Since 16 of the women take both subjects we know that $|E \cap M \cap W| = 16$, and it follows that the number of women who just take Mathematics is $(20 - 16) = 4$, and the number of women who just take Economics is $(45 - 16) = 29$. Similarly, since 24 students take both subjects, and 16 of these are women, it follows that the number of men who take both subjects is $(24 - 16) = 8$. You can see more clearly how we arrived at these figures by looking at the positions of the numbers 16, 4, 29 and 8 in the Venn diagram of Fig 2.15. By the same process, we can deduce that the number of men who just take Economics is $(140 - 8 - 29 - 16) = 87$, the number of men who just take Mathematics is $(50 - 8 - 4 - 16) = 22$, and finally that the number of women who take neither subject is $(60 - 29 - 4 - 16) = 11$.

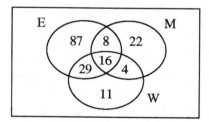

**Fig 2.15**

Now we can add the numbers in the various regions of the Venn diagram to deduce that

$$|E \cup M \cup W| = 87 + 8 + 16 + 29 + 22 + 4 + 11 = 177$$

It follows that the number of men at the party is $200 - 177 = 23$.

## Principle of inclusion and exclusion

To obtain the number of elements in the union of two sets we have to add the numbers of elements in $A$ and in $B$, but this will count elements in the intersection of $A$ and $B$ twice, so we subtract $|A \cap B|$ to get

$$|A \cup B| = |A| + |B| - |A \cap B|$$

For the union of three sets we again count all three sets, but then we will have counted the elements in the intersections of each pair of sets twice, and the elements of $A \cap B \cap C$ three times. So we first subtract the numbers of elements in the intersections of each pair of sets. Since there are three pairs ($A$ and $B$, $A$ and $C$, $B$ and $C$), we will now have removed altogether the elements in $A \cap B \cap C$. So to get the right answer, we must add back in $|A \cap B \cap C|$:

$$|A \cup B \cup C| = |A| + |B| + |C| - |A \cap B| - |B \cap C| - |A \cap C| + |A \cap B \cap C|$$

## ● *Example 33*

Suppose that 80 children went to the fair. There were three attractions: Big Dipper, Ferris Wheel, Dodgems. Thirty of the children went on all three attractions, 50 of them on at least two. Each costs £1 and the total spent was £140. How many children missed out on all three attractions?

SOLUTION
The easiest way to tackle this problem is to consider Fig 2.16, a Venn diagram of three sets, in which each region has been labelled with a letter to stand for the number of children on that particular combination of attractions. Thus, $r$ is the number of children who only went on the Big Dipper.

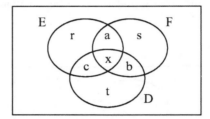

**Fig 2.16**

We know that the number of children who went on all three attractions is 30, so $x$ is 30. From the diagram we can see that the number who went on exactly two attractions is $a + b + c = 50 - 30 = 20$, and the number who went on exactly one is $r + s + t$. Now,

$$(r + s + t) + 2 \times (a + b + c) + 3 \times 30 = 140$$
$$(r + s + t) + 40 + 90 = 140$$
$$r + s + t = 10$$

So, 10 children went on exactly one attraction, which leaves $80 - 10 - 30 - 20 = 20$ children who missed out altogether.

### EXERCISES ON 2.5

1.  Give the inclusion–exclusion rule for four sets.
2.  There are 20 students taking analysis, 30 students taking algebra and 30 students taking statistics. There are 5 students taking analysis and algebra, 10 taking algebra and statistics and 4 taking analysis and statistics. There are 3 students taking all three courses. How many students are taking at least one of the three courses? How many students are just taking analysis?
3.  In a survey of 100 first-year mathematics students, 20 have mothers who have been to university, 15 have fathers who have been to university, and for 70 of

the students, neither of their parents has been to university. For how many of the students have both of their parents been to university?

4. A survey of 500 people showed that 200 of them wore glasses, 50 had false teeth, 300 voted Labour; 40 wore glasses and had false teeth; 10 voted Labour and had false teeth; 150 wore glasses and voted Labour; and 8 voted Labour and wore glasses and had false teeth. How many people voted Labour who did not have glasses or false teeth? How many people did not vote Labour who had false teeth and glasses?

5. Seventy students went to the societies fair. Fifteen of the students joined three societies, and 34 of them joined at least two societies. No student joined more than three societies. It costs £5 to join a society and the total takings were £550. How many students did not join any societies at all?

6. The Discrete Mathematics exam had three questions. Each one of 40 students was able to do at least one question. Ten students were unable to do question 1, 15 students were unable to do question 2, and 20 of the students were unable to do question 3. Only 5 students were able to answer all three questions. How many students answered exactly two questions?

# 2.6 Products and partitions

The student cafeteria offers the following menu:

- main dishes:
  chicken curry
  vegetable lasagne
  haddock
- side dishes:
  rice
  chips
  baked potato

What are the different meal combinations that can be ordered? Let $M$ be the set of main dishes {chicken curry, vegetable lasagne, haddock} and $S$ the set of side dishes {rice, chips, baked potato}. We define the **product** of the two sets $M$ and $S$ to be the set whose elements consist of all possible pairs $(m, s)$ where $m$ is an element of $M$ and $s$ is an element of $S$. So in this example the product of $M$ and $S$, written $M \times S$, gives the set of all possible meal combinations:

$M \times S = \{$(chicken curry, rice), (vegetable lasagne, rice), (haddock, rice), (chicken curry, chips), (vegetable lasagne, chips), (haddock, chips), (chicken curry, baked potato), (vegetable lasagne, baked potato), (haddock, baked potato)$\}$.

Notice that the set $M \times S$ is a set of ordered pairs, called **2-tuples**. We can take the product of any number of sets. The product of $n$ sets will produce a set whose elements are **n-tuples**.

## Example 34

Let $A = \{a, b, c\}, B = \{2, 4\}$ and $C = \{T, F\}$, then

$$A \times B \times C = \{a, 2, T), (a, 2, F), (a, 4, T), (a, 4, F), (b, 2, T), \ldots\}$$

The set $A \times B \times C$ has 12 elements and each element is a 3-tuple.

If we have a set of students, we can define two subsets by putting all the women into one subset, and all the men into the other. Obviously, each student is then in exactly one of the subsets. But if we classified the same set of students by putting into one set all the students who play a musical instrument, and into the other all the students who take regular exercise, some of the students may end up in both subsets, whilst others may end up in neither.

We say that a collection of non-empty sets $Y_1, \ldots, Y_n$ is a **partition** of a set $X$ if the union of the sets is $X$ and the sets are disjoint, so that

$$Y_1 \cup \ldots \cup Y_n = X \text{ and } Y_i \cap Y_j = \emptyset, \text{ for all } j \neq i$$

So if $S = \{\text{students}\}$, $Y_1 = \{\text{women}\}$ and $Y_2 = \{\text{men}\}$, then $Y_1, Y_2$ is a partition of $S$. But if $Y_1 = \{\text{musicians}\}$ and $Y_2 = \{\text{exercisers}\}$, then $Y_1, Y_2$ is not a partition of $S$.

## Example 35

Let $A = \{1, 2, 3, 4, 5, 6, 7, 8\}$, then $\{1,2\}, \{3\}, \{4,6\}, \{5,7,8\}$ is a partition of $A$.

### EXERCISES ON 2.6

1. Suppose that each of the meals in §2.6 could be served with or without salt. Show how you can write the set of all possible meals as a product of three sets.
2. Let $A = \{a, b\}$ and $B = \{l, m, n\}$. What is $A \times B$?
3. If $X \times Y = Y \times X$ what can you say about the sets $X$ and $Y$?
4. List all the partitions of the set $\{1\}$.
5. List all the partitions of the set $\{a, b\}$.
6. List all the partitions of the set $\{1, 2, 3\}$.

## 2.7 Finite and infinite

Most of the sets we have been considering so far have been finite sets, for example $\{2, 4, 6\}$. But the set of natural numbers, $\mathbb{N} = \{1, 2, 3, \ldots\}$, is an infinite set. For finite sets we can easily see if two sets are the same size, by simply counting that the number of elements in each set is the same. Formally, this is equivalent to establishing a one-to-one correspondence between pairs of elements, one from each set.

## Example 36

Let $A = \{1, 2, 3\}$ and $B = \{2, 4, 6\}$. Then $A$ and $B$ are the same size because we can put the elements of $A$ in one-to-one correspondence with the elements of $B$. One of several possible ways to do this is

$$1 \leftrightarrow 2$$
$$2 \leftrightarrow 4$$
$$3 \leftrightarrow 6$$

(there are other possibilities, because the order of the elements in each set does not matter).

This is unnecessarily cumbersome for such small sets, but is useful when the sets are defined in such a way that direct counting is difficult—and essential when the sets are infinite.

Another example of an infinite set is the set of integers, $\mathbb{Z} = \{\ldots, -2, -1, 0, 1, 2, \ldots\}$. We can see that $\mathbb{N}$ is a subset of $\mathbb{Z}$, and that $\mathbb{N}$ is not equal to $\mathbb{Z}$. However, the surprising thing is that we can still put the two sets in one-to-one correspondence with each other. For example,

$$1 \leftrightarrow 0$$
$$2 \leftrightarrow -1$$
$$3 \leftrightarrow 1$$
$$4 \leftrightarrow -2$$
$$5 \leftrightarrow 2$$

and so on.

We can also illustrate the correspondence in Fig 2.17.

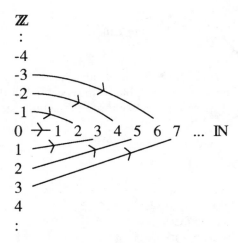

**Fig 2.17**

We call sets that can be put into one-to-one correspondence with the natural numbers **countable.** Countable sets can have all their elements listed one after another.

### ● *Example 37*

Squares of positive integers are countable.

$$
\begin{array}{ccccccc}
1 & 2 & 3 & 4 & 5 & 6 & \ldots \\
\downarrow & \downarrow & \downarrow & \downarrow & \downarrow & \downarrow & \downarrow \\
1 & 4 & 9 & 16 & 25 & 36 & \ldots
\end{array}
$$

What about the set of rational numbers, $\mathbb{Q}$? This set consists of integers and numbers like $\frac{1}{4}$ or $-\frac{2}{7}$ which are ratios of integers. The formal definition is

$$
\mathbb{Q} = \left\{ \frac{a}{b} : a \in \mathbb{Z}, b \in \mathbb{Z} \text{ and } b \neq 0 \right\}
$$

Is $\mathbb{Q}$ countable? We shall prove that $\mathbb{Q}^+$, the set of positive rational numbers, is countable, although the same idea works for $\mathbb{Q}$ itself.

$$
\begin{array}{c|cccccc}
 & \multicolumn{6}{c}{a} \\
 & 1 & 2 & 3 & 4 & \ldots \\
\hline
1 & 1 & 2 & 3 & 4 & \ldots \\
2 & \frac{1}{2} & \frac{2}{2}=1 & \frac{3}{2} & \frac{4}{2} & \ldots \\
3 & \frac{1}{3} & \frac{2}{3} & \frac{3}{3} & \frac{4}{3} & \ldots \\
4 & \frac{1}{4} & \frac{2}{4} & \frac{3}{4} & \frac{4}{4} & \ldots \\
5 & \frac{1}{5} & \frac{2}{5} & \frac{3}{5} & \frac{4}{5} & \ldots \\
6 & \frac{1}{6} & \frac{2}{6} & \frac{3}{6} & \frac{4}{6} & \ldots \\
\vdots & \vdots & \vdots & \vdots & \vdots & \vdots
\end{array}
$$

with $b$ labelling the rows.

**Fig 2.18**

We can put all the elements of $\mathbb{Q}^+$ in a table as shown in Fig 2.18. All the elements of $\mathbb{Q}$ can be included in this table (with repeats), but the table as it stands is not a list. However, if we run snake-like through the table, as shown in Fig 2.19,

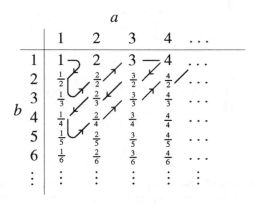

**Fig 2.19**

we can turn it into a list:

$$1, 2, \frac{1}{2}, \frac{1}{3}, \frac{2}{2}, 3, 4, \frac{3}{2}, \frac{2}{3}, \frac{1}{4}, \frac{1}{5}, \ldots$$

Now we can remove all the repeats, to get another list:

$$1, 2, \frac{1}{2}, \frac{1}{3}, 3, 4, \frac{3}{2}, \frac{2}{3}, \frac{1}{4}, \frac{1}{5}, \ldots$$

This list will contain all the elements of $\mathbb{Q}^+$. So $\mathbb{Q}^+$ is countable. What about the real numbers, $\mathbb{R}$? The set $\mathbb{R}$ includes the rational numbers as well as numbers like $\sqrt{2}, \pi, e, \sqrt{3}$ etc. Elements of $\mathbb{R}$ which are not also elements of $\mathbb{Q}$ are called **irrational numbers**. Can we put $\mathbb{R}$ in one-to-one correspondence with the natural numbers?

The technique we shall use to answer this question is nicely illustrated by the following example.

## ● *Example 38*

Three children draw up a table of their likes and dislikes of three varieties of sweets:

|  | Sweets | | |
|---|---|---|---|
| 1 | Chocolate | Humbug | Peppermint |
| Abel | like | like | like |
| Babs | hate | hate | like |
| Chloe | like | like | like |

Suppose that we find a child who disagrees with Abel about Chocolate, disagrees with Babs about Humbug, and disagrees with Chloe about Peppermint. Can this person be $A$, $B$ or $C$? No—this must be a different person, since they disagree with each of the tastes of $A$, $B$ and $C$ at least once.

Now, as so often in mathematics, we can extend this argument for $n$ children and $n$ types of sweet. Suppose there is a child $X$ who disagrees with child 1 about sweet 1, with child 2 about sweet 2, ..., with child $n$ about sweet $n$. Can child $X$ be one of the $n$? No, $X$ must be a different child.

We can now use the same technique to show that all reals between 0 and 1 are not countable. Suppose that the set of reals between 0 and 1 is countable. Write down their decimal representations in a list, where if the decimal terminates we shall end it with an infinite sequence of zeros. Thus,

$0.a_{11}a_{12}a_{13}a_{14}\ldots$

$0.a_{21}a_{22}a_{23}a_{24}\ldots$

$\vdots$

$0.a_{i1}a_{i2}a_{i3}\ldots$

$\vdots$

where $a_{ij}$ denotes the $j$th digit of the $i$th number on the list.

Consider the number

$$b = 0.b_1 b_2 b_3 b_4 \ldots$$

where

$$b_i = \begin{cases} 1 & \text{if } a_{ii} = 9 \\ 9 - a_{ii} & \text{if } a_{ii} = 0, 1, 2, \ldots, 8 \end{cases}$$

Notice that this definition rules out an infinite sequence of nines so as to avoid any ambiguity.

We can see that $b$ is a real number between 0 and 1, and that $b$ is not on our list! This is because $b_1$ is different from $a_{11}$, $b_2$ is different from $a_{22}$, etc. Hence, the set of real numbers between 0 and 1 is *uncountable*. We have shown that a subset of $\mathbb{R}$ is uncountable, and hence that $\mathbb{R}$ is uncountable.

## EXERCISES ON 2.7

1. Show that the union of two countable sets is countable.
2. True or false?
   (a) The intersection of two countable sets is countable.
   (b) The set of all subsets of $\mathbb{N}$ is countable.
   (c) A subset of a finite set is finite.
   (d) If $A$ is a set and $B = \mathcal{P}(A)$ then there is a one-to-one correspondence between elements of $A$ and elements of $B$.
   (e) There is a one-to-one correspondence between $\mathbb{N}$ and the set of all positive even numbers.
3. Let $S$ be the set of all quadratic equations of the form

$$x^2 + mx + n = 0$$

   where $m$ and $n$ are integers. Prove that $S$ is countable.

# 2.8 Paradoxes

The central idea of set theory is that it provides an unambiguous way of classifying things. Given any universal set $\mathcal{U}$, and any definition of a set $S$, it should be possible to say whether any $u \in \mathcal{U}$ is or is not an element of $S$. However, in 1901 Bertrand Russell (1872–1970), an English philosopher and mathematician, shook mathematicians' belief in set theory and the foundations of arithmetic by propounding the following paradox.

## Russell's paradox

Russell defined a set $S$ as

$$S = \{\text{sets which are not elements of themselves}\},$$

and asked the following question: is the set $S$ an element of the set $S$?

If $S$ is an element of the set $S$ then $S$ should satisfy the property that $S$ is not an element of itself. Conversely, if $S$ is not an element of $S$ then it should be in the set

**Fig 2.20** Bertrand Russell

$S$, i.e. $S$ is an element of $S$. So we have a logical contradiction, often called a paradox.

In case you wonder what connection this can possibly have with the real world, we now give a less abstract version of Russell's paradox.

## The barber's paradox

The barber in a certain village shaves everyone who does not shave themselves. So we can define the set

$B =$ {people who do not shave themselves}

and ask the question: does the barber shave himself? If the barber shaves himself, then by the definition of $B$ he is not an element of $B$. But the barber only shaves people in $B$, so he should not shave himself. Conversely, if the barber does not shave himself then he should be in the set $B$, and the barber should shave him. In each case we have a contradiction.

Another paradox of the same kind is

The sentence you are now reading is false.

If the statement is true, then it says that it is false, and if the statement is false, then it says that it is true!

To counter Russell's paradox, a new set theory was developed in 1908 by Ernst Zermelo, a German mathematician who worked on integral equations. Zermelo's theory was later modified by Abraham Fraenkel. Most mathematicians believe that nobody will be able to find any paradoxes within the Zermelo–Fraenkel set theory, but this has not yet been proved!

### EXERCISES ON 2.8

1. Consider the set $C =$ {catalogues which do not list themselves}. Is $C$ an element of itself?

2. An adjective is said to be heterological if it describes a property that it does not possess. For example, the word 'English' is not heterological, but the word 'French' *is* heterological. Is the adjective 'heterological' heterological? (This example is taken from *Elements of Discrete Mathematics* by C. L. Liu, published by McGraw-Hill.)

# 2.9 Projects

## Collections of sets

Let $S_2$ be the set of all quadratic equations of the form

$$x^2 + m_1 x + m_0 = 0$$

where $m_0$ and $m_1$ are integers. One of the earlier exercises asked you to prove that $S$ is countable. Now consider a whole collection of sets, indexed by a positive integer $k$, such that $S_k$ is the set of all polynomial equations of degree $k$ of the form

$$x^k + m_{k-1} x^{k-1} + \ldots + m_1 x + m_0 = 0$$

where $m_0, m_1, \ldots, m_{k-1}$ are integers. Define

$$P = S_2 \cup S_3 \cup \ldots$$

This is a union of an infinite number of sets, each of which has an infinite (but countable) number of elements. Is $P$ countable?

## Bounds on the size of a union of sets

For any two sets $S_1$ and $S_2$, we know that the number of elements in $S_1 \cup S_2$ is less than or equal to the sum of the number of elements in $S_1$ and the number of elements in $S_2$. In symbols,

$$|S_1 \cup S_2| \leq |S_1| + |S_2|$$

If you draw a Venn diagram, it is easy to convince yourself that the same holds for three sets, that is

$$|S_1 \cup S_2 \cup S_3| \leq |S_1| + |S_2| + |S_3|$$

Can you prove this?

Now, try to prove the generalisation of this result, i.e. for any sets $S_1, S_2, \ldots, S_n$,

$$|S_1 \cup \ldots \cup S_n| \leq |S_1| + \ldots + |S_n|$$

You will need to use one of the methods of proof which we discussed in Chapter 1. It is also easy to see that

$$|S_1 \cup S_2| \geq |S_1|$$

and similarly that

$$|S_1 \cup S_2| \geq |S_2|$$

One way to combine the last two statements into one is to write

$$|S_1 \cup S_2| \geq \max|S_i|$$

Try to prove that, for any positive integer $n$,

$$|S_1 \cup \ldots \cup S_n| \geq \max|S_i|$$

Combining the last two results, we can give the following lower and upper bounds on the size of a union of sets:

$$\max|S_i| \leq |S_1 \cup \ldots \cup S_n| \leq |S_1| + \ldots + |S_n|$$

In applications, the actual size of $S_1 \cup \ldots \cup S_n$ is often much bigger than the lower bound, and much smaller than the upper bound. This suggests the question—can we find better (narrower) bounds? In fact, we cannot. Try to prove this, again using one of the general methods of proof which we discussed in Chapter 1.

# *Summary*

A **set** is a collection of specified objects. The objects which make up a set are called **elements** of the set. These elements are displayed between **braces** {}. The braces are read as: the set whose elements are.

The symbol $\in$ represents 'is an element of' and $\notin$ represents 'is not an element of'.

The set $A$ is a **subset** of the set $B$ if every element of $A$ is also an element of $B$. This is written as $A \subseteq B$.

Two sets are equal if they have exactly the same elements. If two sets $A$ and $B$ are equal, then

$$A \subseteq B \text{ and } B \subseteq A$$

We denote the number of elements in a set $X$ by $|X|$.

The **union of A and B** is $A \cup B = \{x : x \in A \text{ or } x \in B\}$. The **intersection of $A$ and $B$** is $A \cap B = \{x : x \in A \text{ and } x \in B\}$. The **difference between $A$ and $B$** is $A \backslash B = \{x : x \in A \text{ and } x \notin B\}$.

$\emptyset$ denotes the **empty set** and $\mathcal{U}$ denotes the **universal set**. The set of elements not in $A$ is called the **complement of $A$**, and is written as $A^c$.

Associativity:

$$(A \cup B) \cup C = A \cup (B \cup C)$$
$$(A \cap B) \cap C = A \cap (B \cap C)$$

Distributivity:

$$A \cap (B \cup C) = (A \cap B) \cup (A \cap C)$$
$$A \cup (B \cap C) = (A \cup B) \cap (A \cup C)$$

De Morgan's laws:

$$(A \cup B)^c = A^c \cap B^c$$
$$(A \cap B)^c = A^c \cup B^c$$

The set of all subsets of a set $X$ is called the **power set** of $X$, denoted by $\mathcal{P}(X)$. The power set $\mathcal{P}(X)$ has $2^{|X|}$ elements.

The inclusion–exclusion rule for two sets is

$$|A \cup B| = |A| + |B| - |A \cap B|$$

and the corresponding rule for three sets is

$$|A \cup B \cup C| = |A| + |B| + |C| - |A \cap B| - |B \cap C| - |A \cap C| + |A \cap B \cap C|$$

The **product** of two sets $M$ and $S$ is the set whose elements consist of all possible pairs $(m, s)$ where $m$ is an element of $M$ and $s$ is an element of $S$. The elements are called 2-tuples.

A collection of non-empty sets $Y_1, \ldots, Y_n$ is a **partition** of a set $X$ if the union of the sets is $X$ and the sets are disjoint, so that

$$Y_1 \cup \ldots \cup Y_n = X \text{ and } Y_i \cap Y_j = \emptyset \text{ for all } j \neq i$$

# 3 • Relations and Functions

Classification is an important part of mathematics. By classifying problems in various ways, we are able to develop general methods for solving them. We have seen in Chapter 2 on sets how we can group very general objects together. For example, if $P$ is the set of polynomial equations, we can partition $P$ into subsets consisting of all linear equations, all quadratic equations, and so on. In order to classify mathematical objects, we need to be able to talk about the relationships which our objects have to one another. Formally, *relations* in mathematics describe connections between different elements of the same set, whereas *functions* describe connections between two different sets. Imagine a set $A$ of people,

$A = \{$Amanda, Barbi, Bryony, Deborah, Holly, John, Jonathan,
    Kate, Meg, Peter$\}$

and a set $B$ of ages,

$B = \{1, 2, 3, \ldots\}$

The phrase 'is younger than' defines a *relation* between any two elements of $A$. For example, Deborah is younger than Amanda, but Peter is not younger than Jonathan. On the other hand, if we actually want to know how old the people in $A$ are, we need to make a connection between elements of $A$ and elements of $B$—so 'age' is a *function* which operates between the sets $A$ and $B$. For example, Jonathan is age 1 and Holly is age 12.

In this chapter, we first look at the definition and basic properties of a *relation*. We then consider two important types of relation, equivalence relations and partial orderings. Next, we define a *function* and consider several different ways of visualising what a function does. We show how functions can be combined using the idea of a *composition* of functions, and consider the important properties of one-one, onto and invertible. We end with a discussion of the *pigeonhole principle*, which allows us to solve all kinds of seemingly unrelated problems.

## 3.1 Relations

When we hear the word 'relations' we usually think about family relationships, and a family tree (Fig 3.1) gives a nice introduction to the more general ideas of relations in mathematics.

### Symmetric relationships

From the family tree you can see that Penny is Lisa's sister and Lisa is Penny's sister. Is it always true that if A is B's sister, then B is A's sister?

Lisa is Jack's sister but Jack is not Lisa's sister.

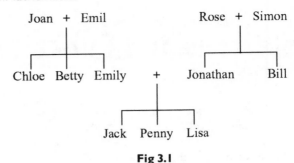

**Fig 3.1**

The relationship 'is a sibling of' (a sibling is either a brother or a sister) *does* always work both ways.

If A is a sibling of B then B is a sibling of A for all possible A and B.

Relationships with this property are said to be *symmetric*.

In the definition of a symmetric relationship, it is not enough to find one example where the relationship works in both ways. A relationship is only symmetric if the symmetry holds in all cases. If A is a sister of B, then B may well be a sister of A, but this is not always the case, so 'is a sister of' is *not* a symmetric relationship.
We can think about relations other than those between people. For example, if $x$ and $y$ are integers, the statement

$x$ is greater than $y$

is a relation between $x$ and $y$. Does this imply that $y$ is greater than $x$? Obviously not—so 'is greater than' is *not* a symmetric relation on the integers.

In deciding whether a given relation is symmetric, or what other properties it might have, we need to be clear about the set on which we are considering the relation. For example, Amanda Chetwynd has two sisters and no brothers, so *within her family* the relation 'is a sister of' *is* symmetric. Our formal definition of a symmetric relation is therefore as follows.

A relation **R** on a set $S$ is said to be **symmetric** if

$a\mathbf{R}b \Rightarrow b\mathbf{R}a$ for all elements $a, b$ in $S$

## Transitive relationships

To return to our family tree, can you find three people, A, B and C, so that A is the same sex as B, B is the same sex as C, and A is the same sex as C? Does the third 'is the same sex as' follow automatically from the first two?

Can you also find three people so that A is the son of B and B is the son of C? Does it then follow that A will be the son of C?

We say that 'is the same sex as' is a *transitive* relationship whereas 'is the son of' is *not* a transitive relationship.

Suppose that $x, y$ and $z$ are integers, and that $x > y$ and $y > z$. What can be said about the relation between $x$ and $z$? Since 'is greater than' is a transitive relation, we can say that $x > z$.

A relation **R** on a set $S$ is said to be **transitive** if

$a\mathbf{R}b$ and $b\mathbf{R}c \Rightarrow a\mathbf{R}c$ for all elements $a, b, c$ in $S$

## Reflexive relationships

There is a third property that a relationship may or may not have. For example, consider the relation 'is the same sex as'. Is it true that Lucy is the same sex as Lucy? Is it true for any person A that A is the same sex as A? A relationship with this property is called *reflexive*.

Is the relationship 'is the mother of' reflexive?

As before, we can apply the same ideas to objects other than people. For example, the relation 'is less than or equal to' is reflexive on the integers, since any integer is less than or equal to itself. A relation **R** on a set $S$ is said to be **reflexive** if

$a\mathbf{R}a$ for all elements $a$ in $S$

### EXERCISES ON 3.1

1.  Which of the following relationships are symmetric?
    (a) has the same parents as
    (b) is the father of
    (c) is married to
    (d) is an aunt of
2.  Here are some more relations and the objects that they compare. Decide which of the relations are symmetric. For those that are not symmetric, give a counterexample.
    (a) 'is less than' for integers
    (b) 'is the opposite of' for words in English
    (c) 'is the same distance from the origin as' for points on the plane
    (d) 'is a factor of' for integers
    (e) 'has the same remainder on division by 3 as' for integers
    (f) 'is perpendicular to' for lines in the plane
    (g) 'is parallel to' for lines in the plane
3.  For the relations given above in Exercise 2, decide which ones are transitive.
4.  For the relations given above in Exercise 2, decide which ones are reflexive.
5.  Give an example of a relation which is
    (a) reflexive, symmetric, but not transitive
    (b) reflexive, transitive, but not symmetric
    (c) symmetric, transitive, but not reflexive

## 3.2 Equivalence relations

Relations which possess all three properties of reflexivity, symmetry and transitivity are called *equivalence relations*. We use the notation $a \equiv b \pmod{k}$ for $a$ is congruent to $b$ *modulo* $k$. This means that $a$ and $b$ have the same remainder on division by $k$.

### ● *Example I*

Show that the relation 'is congruent *modulo* 4 to' on the the set of integers $\{0, 1, \ldots, 10\}$ is reflexive, symmetric and transitive.

SOLUTION

We first prove that $a \equiv b \pmod 4$ is symmetric. To do this, we need to show that

$$a \equiv b \pmod 4 \Rightarrow b \equiv a \pmod 4$$

If $a \equiv b \pmod 4$, then $a = 4n + c$ and $b = 4m + c$ for some integers $m$ and $n$, with $c = 0, 1, 2$ or $3$. Then $a - b = 4(n - m)$ is an integer multiple of 4. We say that 4 divides $a - b$ with no remainder, and write this as $4|(a - b)$. Then,

$$
\begin{aligned}
a \equiv b \pmod 4 \ &\Rightarrow 4|(a - b) \\
&\Rightarrow 4|(b - a) \\
&\Rightarrow b \equiv a \pmod 4
\end{aligned}
$$

To prove that $a \equiv b \pmod 4$ is transitive, we need to show that $a \equiv b \pmod 4$ and $b \equiv c \pmod 4 \Rightarrow a \equiv c \pmod 4$.

$$
\begin{aligned}
a \equiv b \pmod 4 \text{ and } b \equiv c \pmod 4 \ &\Rightarrow 4|(a - b) \text{ and } 4|(b - c) \\
&\Rightarrow a - b = 4k \text{ and } b - c = 4l \\
&\quad\ \text{for some integers } k, l \\
&\Rightarrow a - b + b - c = 4k + 4l \\
&\Rightarrow a - c = 4(k + l) \\
&\Rightarrow 4|(a - c) \\
&\Rightarrow a \equiv c \pmod 4
\end{aligned}
$$

To prove that $a \equiv b \pmod 4$ is reflexive, we need to show that $a \equiv a \pmod 4$.

$$
\begin{aligned}
a - a = 0 \times 4 &\Rightarrow 4|(a - a) \\
&\Rightarrow a \equiv a \pmod 4
\end{aligned}
$$

Hence, the relation $a \equiv b \pmod 4$ is reflexive, symmetric and transitive.

Consider the integers between 0 and 10. Which ones are related, in the sense that they are equivalent to each other (mod 4)? For example, the integers 0, 4 and 8 are all related to each other in this sense, since $0 \equiv 4 \pmod 4$, $0 \equiv 8 \pmod 4$ and $4 \equiv 8 \pmod 4$.

Choose an integer other than 0, 4 or 8. Find all the integers which are congruent modulo 4 to your chosen integer. Keep doing this until you have accounted for all the integers up to 10.

You will have found that the relation has partitioned the numbers 0 to 10 into sets $\{0, 4, 8\}$, $\{1, 5, 9\}$, $\{2, 6, 10\}$ and $\{3, 7\}$. We call these sets *classes*. In each class, every element is related to all other elements in the class, and not to any element of another class. The relation partitions the integers into classes of related integers, the relation in this case being that all the integers in a given class have the same remainder when divided by 4. The classes are sometimes called *residue classes*. We will use the more common name of *equivalence classes*.

# ● *Example 2*

Using the same set of integers 0 to 10, show how the relation $a \equiv b$ (mod 5) divides the set into classes. Are the classes again distinct?

SOLUTION
The relation $a \equiv b$ (mod 5) partitions the set $\{0, 1, 2, 3, 4, 5, 6, 7, 8, 9, 10\}$ into distinct classes $\{0, 5, 10\}$, $\{1, 6\}$, $\{2, 7\}$, $\{3, 8\}$ and $\{4, 9\}$.

Any relation on a given set $S$ which is symmetric, transitive and reflexive is called an **equivalence relation**. An equivalence relation always partitions $S$ into classes, called **equivalence classes.**

# ● *Example 3*

The relation *equals* is an equivalence relation on any set. The equivalence classes are the individual elements of the set.

# ● *Example 4*

We can prove that $a \equiv b$ (mod $n$) is an equivalence relation on the integers, for any given value of $n$.

SOLUTION
To prove that congruence is symmetric, we show that $a \equiv b$ (mod $n$) $\Rightarrow b \equiv a$ (mod $n$):

$$a \equiv b \text{ (mod } n) \Rightarrow n|(a - b)$$
$$\Rightarrow n|(b - a)$$
$$\Rightarrow b \equiv a \text{ (mod } n)$$

To prove that congruence is transitive, we show that $a \equiv b$ (mod $n$) and $b \equiv c$ (mod $n$) $\Rightarrow a \equiv c$ (mod $n$):

$$a \equiv b \text{ (mod n)} \text{ and } b \equiv c \text{ (mod n)} \Rightarrow n|(a - b) \text{ and } n|(b - c)$$
$$\Rightarrow a - b = kn \text{ and } b - c = ln$$
$$\text{for some integers } k, l$$
$$\Rightarrow a - b + b - c = kn + ln$$
$$\Rightarrow a - c = (k + l)n$$
$$\Rightarrow n|(a - c)$$
$$\Rightarrow a \equiv c \text{ (mod } n)$$

To prove that congruence is reflexive, we show that $a \equiv a$ (mod $n$):

$$a - a = 0n \Rightarrow n|(a - a)$$
$$\Rightarrow a \equiv a(\text{mod } n)$$

Hence, the relation $a \equiv b$ (mod $n$) has all three properties and the relation partitions the integers into related classes. Any two integers are in the same class if they have the same remainder when divided by $n$.

## ● *Example 5*

Show that the relation $(x, y)\mathbf{R}(a, b) \Leftrightarrow x^2 + y^2 = a^2 + b^2$ is an equivalence relation on the plane, and describe the equivalence classes.

SOLUTION
The following argument shows that $\mathbf{R}$ is symmetric:

$$(x, y)\mathbf{R}(a, b) \Rightarrow x^2 + y^2 = a^2 + b^2$$
$$\Rightarrow a^2 + b^2 = x^2 + y^2$$
$$\Rightarrow (a, b)\mathbf{R}(x, y)$$

Hence, $\mathbf{R}$ is symmetric.
The following argument shows that $\mathbf{R}$ is transitive:

$$(x, y)\mathbf{R}(a, b) \text{ and } (a, b)\mathbf{R}(c, d) \Rightarrow x^2 + y^2 = a^2 + b^2 \text{ and } a^2 + b^2 = c^2 + d^2$$
$$\Rightarrow x^2 + y^2 = c^2 + d^2$$
$$\Rightarrow (x, y)\mathbf{R}(c, d)$$

Hence, $\mathbf{R}$ is transitive.
The following argument shows that $\mathbf{R}$ is reflexive:

$$(x, y)\mathbf{R}(x, y) \Leftrightarrow x^2 + y^2 = x^2 + y^2$$

Hence, $\mathbf{R}$ is reflexive.
Now, for any point $(x, y)$ the sum $x^2 + y^2$ is the square of its distance from the origin. The equivalence classes are therefore the sets of points in the plane which have the same distance from the origin. Thus, the equivalence classes are concentric circles centred on the origin.

## EXERCISES ON 3.2

1. Which of the following relations are equivalence relations?
   (a) 'equals' on the set $\mathbb{N}$
   (b) 'greater than' on the set $\mathbf{Z}$
   (c) 'sister of' on the set of people in the world
   (d) 'perpendicular to' on the set of lines in the plane
   (e) 'parallel to' on the set of lines in the plane
2. Prove that the following relations are equivalence relations and describe the equivalence classes:
   (a) the relation $m\mathbf{R}n \Leftrightarrow m^2 - n^2$ is divisible by 3 on the set $\mathbf{Z}$
   (b) the relation $(x, y)\mathbf{R}(a, b) \Leftrightarrow x + y = a + b$ on the plane
   (c) the relation $(x, y)\mathbf{R}(a, b) \Leftrightarrow \lfloor x \rfloor + \lfloor y \rfloor = \lfloor a \rfloor + \lfloor b \rfloor$ on the plane ($\lfloor x \rfloor$ means the greatest integer $\leq x$)
   (d) the relation $(x, y)\mathbf{R}(a, b) \Leftrightarrow xb = ay$ on the set $\mathbf{Z} \times \mathbf{Z}$
3. For the equivalence relation

$$m\mathbf{R}n \Leftrightarrow mn \text{ is a square on the set of positive integers}$$

find the equivalence classes of
(a) 1
(b) 2
(c) 5
(d) $p$, where $p$ is a prime number
4. Describe the equivalence class of the point $(1\frac{1}{4}, 2\frac{1}{2})$ and the partitioning of the plane by the following equivalence relations:
(a) $(x, y)\mathbf{R}(a, b)$ means $x^2 + y^2 = a^2 + b^2$
(b) $(x, y)\mathbf{R}(a, b)$ means $x + y = a + b$
(c) $(x, y)\mathbf{R}(a, b)$ means $\lfloor x \rfloor = \lfloor a \rfloor$ and $\lfloor y \rfloor = \lfloor b \rfloor$
5. What is wrong with the following 'proof' that if a relation is symmetric and transitive, then it must be reflexive? If $a\mathbf{R}b$ then $b\mathbf{R}a$ by symmetry. If $a\mathbf{R}b$ and $b\mathbf{R}a$ then $a\mathbf{R}a$ by transitivity. Hence, $a\mathbf{R}a$ for all $a$. Can you find a counter-example?

# 3.3 Partial orders

We have seen that the properties of reflexivity, symmetry and transitivity define a special kind of relation, called an equivalence relation. We now introduce a new property, antisymmetry.

## Antisymmetric relationships

The relation 'less than or equal to' is not symmetric for integers. For example, 3 is less than or equal to 4 but 4 is not less than or equal to 3. However, if two integers $x$ and $y$ satisfy $x \leq y$ and $y \leq x$, then it follows that $x = y$. Relationships which are only symmetric when the two elements are the same are called *antisymmetric*. The relationship 'less than or equal to' is antisymmetric on the integers.

A relation $R$ on a set $S$ is said to be **antisymmetric** if

$$a\mathbf{R}b \text{ and } b\mathbf{R}a \Rightarrow a = b \text{ for all elements } a, b \text{ in } S$$

We now consider relations which have all three properties of reflexivity, antisymmetry and transitivity.

## ● *Example 6*

Show that the relation 'less than or equal to' on the integers is reflexive, antisymmetric and transitive.

SOLUTION
$a \leq a$ for all $a \in \mathbf{Z}$, hence $\leq$ is reflexive. If $a \leq b$ and $b \leq a$, then $a = b$, hence $\leq$ is antisymmetric. If $a \leq b$ and $b \leq c$ then $a \leq c$, hence $\leq$ is transitive.

Relations which are reflexive, antisymmetric and transitive on a set are called **partial orders**. The set is then called a **partially ordered set** or **poset**. Another example of a partial order is the relation 'greater than or equal to'.

## Example 7

Let $\mathcal{P}(S)$ be the power set of $S = \{a, b, c\}$. Show that the subset relation, $\subseteq$, defines a partial order on $\mathcal{P}(S)$.

SOLUTION

$\mathcal{P}(S) = \{\emptyset, \{a\}, \{b\}, \{c\}, \{a, b\}, \{a, c\}, \{b, c\}, \{a, b, c\}\}$. Every set is included in itself, hence $\subseteq$ is reflexive. Let $X$ and $Y$ be any members of $\mathcal{P}(S)$. If $X \subseteq Y$ and $Y \subseteq X$, then $X = Y$, hence $\subseteq$ is antisymmetric. Let $X$, $Y$ and $Z$ be any three members of $\mathcal{P}(S)$. If $X \subseteq Y$ and $Y \subseteq Z$, then $X \subseteq Z$, hence $\subseteq$ is transitive.

### EXERCISES ON 3.3

1. Show that 'divides' on the set of integers is a partial order.
2. Which of the following properties does **R** possess: reflexive, symmetric, antisymmetric, transitive?
   (a) $m\mathbf{R}n$ if and only if $m - n$ is divisible by 2 or 3 where $m, n \in \mathbf{Z}$.
   (b) $m\mathbf{R}n$ if and only if $m - n$ is divisible by 2 and 3 where $m, n \in \mathbf{Z}$.
   (c) $m\mathbf{R}n$ if and only if $m \geq n$ where $m, n \in \mathbf{Z}$.
   (d) $m\mathbf{R}n$ if and only if $m < n + 1$ where $m, n \in \mathbb{R}$.
   (e) $m\mathbf{R}n$ if and only if $m < n - 1$ where $m, n \in \mathbb{R}$.
3. Is the relation 'is less than' antisymmetric for the integers?
4. Is the relation 'contains the same vowels as' a partial order on the set of words in the *Oxford English Dictionary*?

## 3.4 Diagrams of relations

In this section we shall show how we can represent a relation **R** on a set $S$ using a diagram and how it is then easy to check the properties of reflexivity, symmetry and transitivity.

A **diagram** of a relation **R** on a set $S$ is a picture consisting of the individual elements of $S$, with an arrow pointing from $a$ to $b$ if and only if $a\mathbf{R}b$. When you try to construct such a diagram, you will often find that you need to be careful how to lay out the individual elements of $S$ on the page to avoid a confusing jumble of arrows—this is really just a process of trial and error, although practice helps!

## Example 8

Consider the relation 'has the same number of prime factors as' on the set $S = \{2, 3, \ldots, 8\}$. In Fig 3.2, we first arranged the prime numbers 2, 3, 5 and 7 in one group, since we know that they all have one prime factor and so are related to each other. This left us to consider the even numbers 4, 6 and 8. Since $4 = 2 \times 2$ and $6 = 2 \times 3$, they both have two prime factors and we placed them beside each other, leaving $8 = 2 \times 2 \times 2$ as the only element of $S$ with three prime factors.

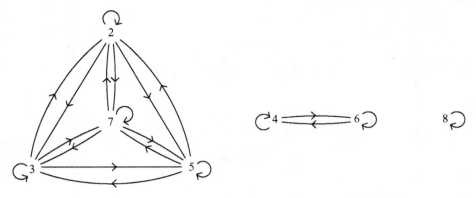

**Fig 3.2**

## Example 9

Draw the diagram for the relation $\leq$ on the set $\{1, 2, 3, 4\}$.

SOLUTION
The solution is shown in Fig 3.3.

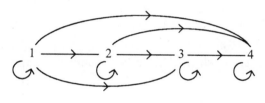

**Fig 3.3**

From these two examples we can see how the properties of reflexivity, symmetry and transitivity are displayed. If a relation is reflexive then in its diagram every element of $S$ will have an arrow pointing to itself. If a relation is symmetric, then whenever there is an arrow from $a$ to $b$ there must also be an arrow from $b$ to $a$. Transitivity needs more careful checking. For any two arrows of the form $a$ to $b$ and $b$ to $c$ there must also be an arrow from $a$ to $c$.

### EXERCISES ON 3.4

1. Draw diagrams for the following relations.
   (a) 'equals' on the set $\{1, 2, 3, 4\}$
   (b) 'divides' on the set $\{1, 2, 3, 4, 5, 6\}$
   (c) 'congruent *modulo* 3 to' on the set $\{0, 1, 2, 3, 4, 5\}$
2. Which of the relations above in Exercise 1 are (a) reflexive, (b) symmetric, (c) antisymmetric, (d) transitive?
3. Figure 3.4 shows some diagrams of relations. For each relation, decide which of the following properties it has:
   (a) reflexivity
   (b) symmetry

(c)  antisymmetry
(d)  transitivity

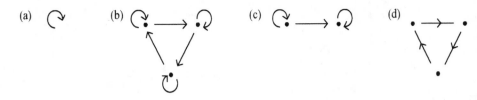

**Fig 3.4**

# 3.5  Functions

In mathematics, a *function* is a rule for transforming a member of one set, *A* say, to a unique member of another set, *B* say. There are several different ways to think about this intuitively.

### Functions as graphs

You are probably familiar with '*y* against *x*' graphs, as in the following example.

### ● *Example 10*

Figure 3.5 shows a graph of $y = x^2$, for *x* lying between $-4$ and $+4$. This corresponds to a *function* which takes a real number, *x*, and transforms it to.

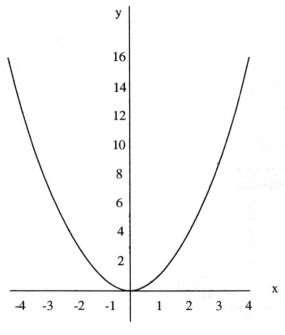

**Fig 3.5**

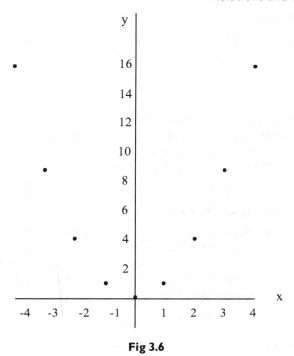

**Fig 3.6**

another real number, $x^2$. In this case, the set $A$ is the interval from $-4$ to $+4$, and the set $B$ is the interval from $0$ to $16$. We could have defined $A$ to be any other interval, or a set of integers. For example, Fig 3.6 shows a graph of $y = x^2$ when $x$ is only allowed to be an integer between $-4$ and $+4$.

Notice that the distinction between $x$ and $y$ matters. The diagram qualifies as a function because, for any $x \in A$ the rule $y = x^2$ gives you a *unique* answer $y$. As you can see, this does not work in reverse. Given a value for $y$, say $y = 4$, there are *two* possible values of $x$, namely $x = -2$ and $x = +2$.

Mathematicians do not mind if two different questions have the same answer, but they insist that a question cannot have two different answers!

## Functions as tables

We are all used to looking up information in tables.

## Example 11

Here is a table of postage rates for second-class letters up to 200 g within the UK.

| Weight not over | Cost |
|---|---|
| 60 g | 19p |
| 100 g | 29p |
| 150 g | 36p |
| 200 g | 43p |

Given any weight of letter up to 200 grams, the table gives a unique answer to the question—how much does it cost to post this letter to a destination within the UK? Notice again that the table does *not* give a unique answer in the reverse direction. If I tell you the cost of posting a letter, you cannot answer the question—what is the weight of this letter? (Although you can give a range of possible answers!)

## Functions as input–output systems

When we enter a number into a pocket calculator, and press one of its keys, for example the $+/-$ key, something predictable happens. The key is acting on the *input* $x$ to produce the *output* $-x$. Suppose you had an enormously complicated calculator with a huge number of keys, but that each key had been painted black. How might you find out what each key does?

## ● *Example 12*

Suppose that when I enter the number 3 and press a particular key, out comes the number 9.

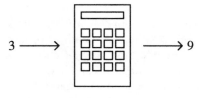

**Fig 3.7**

What do you guess the key is doing? We can try a different input to help confirm the guess. If we enter $-2$ and the output is 4, we might guess that the key is squaring the input.

In the above example, the experiments described do not *prove* that the key is squaring the input. What would you conclude if you entered 4 and the output was 10?

To describe the key in the above example properly, we use the letter $f$ to denote the key, and describe what happens to a general input $x$. Thus, if $f$ is the 'square' function, we say

$$x \to \boxed{f} \to x^2$$

or, slightly more compactly,

$$f(x) = x^2$$

and if $f$ is the 'add 6' function,

$$x \rightarrow \boxed{f} \rightarrow x + 6$$

or

$$f(x) = x + 6$$

## Functions as mappings from one set to another

This provides the most general view of what a function is.

## Example 13

The 26 letters of the English alphabet can be classified as *vowels* or *consonants*. Let $L$ denote the set of letters, $V$ the set of vowels and $C$ the set of consonants. Then

$$V = \{a, e, i, o, u\}$$

and

$$C = \{b, c, d, f, g, h, j, k, l, m, n, p, q, r, s, t, v, w, x, y, z\}$$

We can define a function, $f$ say, which takes any letter of the alphabet and tells us whether it is a vowel or a consonant. So, if $T$ (for 'type') denotes the set with two elements,

$$T = \{\text{vowel}, \text{consonant}\}$$

then $f$ is a function from $L$ to $T$ which, given any letter $x$, takes the value 'vowel' if $x \in V$ and the value 'consonant' if $x \in C$. To indicate that $f$ transforms each element of $L$ to an element of $T$ we write

$$f : L \rightarrow T$$

To show what $f$ does, we write

$$f(x) = \begin{cases} \text{vowel} & : \quad x \in V \\ \text{consonant} & : \quad x \in C \end{cases}$$

## Example 14

Here are two similar functions, presented as mappings from one set to another:

- $f : \mathbb{R} \rightarrow \mathbb{R}$ where $f(x) = x^2$
- $g : \mathbb{Z} \rightarrow \mathbb{Z}$ where $g(x) = x^2$

Notice that the functions in Example 14 are *not* the same function because the second function is restricted to integer values.

## Example 15

Let $w$ denote the weight of a letter, in grams. Then $w$ is a positive real number, which we write as $w \in \mathbb{R}^+$. Let $p$ be the cost of posting the letter, in pence. Then $p$ is a positive integer, $p \in \mathbb{N}$, so we can think of the cost of posting a letter as a function $f : \mathbb{R}^+ \rightarrow \mathbb{N}$. The table in Example 11 gives us some information about $f$, which we could also express as a rather cumbersome formula:

$$f(w) = \begin{cases} 19 & : & w \le 60 \\ 29 & : & 60 < w \le 100 \\ 36 & : & 100 < w \le 150 \\ 43 & : & 150 < w \le 200 \end{cases}$$

Notice that this is not a complete definition of $f$, since the formula does not define the value of $f(w)$ if $w > 200$. It *is* a complete definition of a function $f : I \to \mathbb{N}$, where $I = \{w : 0 \le w \le 200\}$, because it does define the value of $f(w)$ for every $w$ in $I$.

In some examples, a diagram may help you to understand what a function does.

### Example 16

Fig 3.8 is a diagram of a function whose inputs are $a, b, c$ and for which

$$f(a) = 1$$
$$f(b) = 1$$
$$f(c) = 2$$

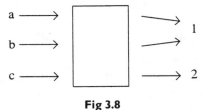

**Fig 3.8**

### Example 17

Fig 3.9 shows a mapping which is *not* a function, because the input $a$ goes to two different things, namely 1 or 2.

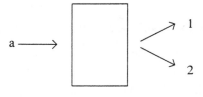

**Fig 3.9**

### Example 18

Which of these are functions?

People $\to$ NHS number

People $\to$ car registration number

SOLUTION
The first is a function. Since a person may own more than one car, the second is not a function.

## Example 19

A material shop sells clothes patterns which specify the amount of material required in metres. The shop still uses an old yard ruler calibrated in inches. The shop needs a way of converting metres to inches.

SOLUTION
The shop needs a function which will take an input in metres and output a measurement in inches. Since 1 metre is approximately equal to 39 inches our function $f$ takes $x$ in metres and outputs $f(x) = 39x$ inches.

## Example 20

A telephone company charges its customers a standard charge for use of the telephone, but has different rates for call units depending on number of units used. Write a function to express the following call charges. The standard charge is £12. For less than 100 units, the rate is £0.05 per unit. For between 100 and 499 units, the rate is £0.04 per unit. For over 500 units the rate is £0.03 per unit.

SOLUTION
$f$ is a function from usage to cost. It is given by

$$f(x) = \begin{cases} 12 + 0.05x & : & 0 \le x < 100 \\ 12 + 0.04x & : & 100 \le x < 500 \\ 12 + 0.03x & : & x \ge 500 \end{cases}$$

Notice how the formula avoids any ambiguity which there might be in the verbal description. For example, we have interpreted the verbal description to mean that if a customer uses 200 units, every one of those units is charged at £0.04. An alternative interpretation might be that the *first* 100 units would charged at £0.05 each, and the remaining 100 at £0.04 each. A formula to express this alternative interpretation is

$$f(x) = \begin{cases} 12 + 0.05x & : & 0 \le x < 100 \\ 17 + 0.04(x - 100) & : & 100 \le x < 500 \\ 33 + 0.03(x - 500) & : & x \ge 500 \end{cases}$$

## Example 21

A function $f$ from the integers to the natural numbers which takes an integer and outputs its modulus plus 1 can be written as

$f : \mathbb{Z} \to \mathbb{N}$ where $f(x) = |x| + 1$

## Example 22

A function $f$ from the integers to the integers which takes an integer and squares it can be written as

$f : \mathbb{Z} \to \mathbb{Z}$ where $f(x) = x^2$

### ● *Example 23*

A function $f$ from the integers to the natural numbers which takes an integer, squares it and adds the result to itself plus 1 can be written as

$$f: \mathbb{Z} \to \mathbb{N} \text{ where } f(x) = x^2 + x + 1$$

We can now give our formal definition of a function. A **function** $f$ from a set $A$ to a set $B$ is a rule which associates with each member of $A$ a unique member of $B$. We write $f: A \to B$.

We call $A$ the **domain** of the function and $B$ the **codomain**. The subset of $B$ which is actually used by the function is called the **range**, i.e. $\{f(x) \in B : x \in A\}$. Notice that the domain of a function is uniquely defined, because the function must do something to every element of the domain. However, the codomain is not uniquely defined. If $B$ is a codomain of a function $f$, then so is any other set $C$ for which $B \subseteq C$. This asymmetry is the formal expression of the fact that, in mathematics, we must always be prepared to specify precisely what questions we are asking, but we may not know in advance precisely what answers we expect to get! If we *can* specify precisely what answers we will get, then we can define the codomain of our function to be its range. This is the most informative specification possible.

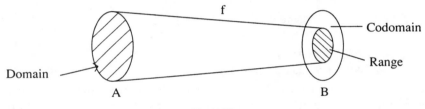

**Fig 3.10**

### ● *Example 24*

In the function

$$f: \mathbb{Z} \to \mathbb{N} \text{ where } f(x) = x^2 + 2$$

the domain of the function is $\mathbb{Z}$, the codomain is $\mathbb{N}$ and the range is $\{2, 3, 6, 11, 18, \ldots\}$.

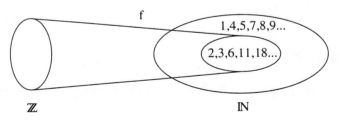

**Fig 3.11**

**EXERCISES ON 3.5**

1. Which of the following mappings are functions?
   (a) $f: \mathbb{Z} \to \mathbb{N}$ where $f(x) = -1$
   (b) $f: \mathbb{Z} \to \mathbb{Z}$ where $f(x) = -x$
   (c) $f: \mathbb{N} \to \mathbb{N}$ where $f(x) = x - 2$
   (d) $f: \mathbb{Z} \to \mathbb{Z}$ where $f(0) = 0$ and $f(x) = x - 1$
   (e) $f: A \to \mathbb{Z}$ where $A = \{1, 2, 3\}$, $f(1) = 0$, $f(2) = 1$, $f(3) = 0$

2. Give the domain, range and codomain of the following functions.
   (a) $f: \mathbb{Z} \to \mathbb{N}$ where $f(x) = 3$
   (b) $f: \mathbb{Z} \to \mathbb{Z}$ where $f(x) = x$
   (c) $f: \mathbb{N} \to \mathbb{Z}$ where $f(x) = x - 2$
   (d) $f: \mathbb{Z} \to \mathbb{Z}$ where $f(x) = |x|$

3. Define a function to convert temperature measured in degrees Celsius to temperature measured in degrees Fahrenheit. What are the domain and codomain of your function?

4. Define a function which describes the charges of a soft drink delivery company. The company charges £2 delivery on orders under 5 bottles and £1 delivery on orders between 5 and 10 bottles, whereas orders over 10 bottles have free delivery. The bottles cost £1 each unless at least 20 are ordered, in which case the cost per bottle is £0.80. Would anybody ever order 19 bottles?

5. Discuss the sense in which you can interpret the phone book as a function. What is the domain? What is the codomain? Can you specify the range?

# 3.6 One-one and onto

Consider the following two functions:

$$f: \mathbb{N} \to \mathbb{N} \text{ where } f(x) = x^2$$

and

$$h: \mathbb{Z} \to \{\mathbb{N} \cup 0\} \text{ where } h(x) = |x|$$

The function $f$ always has a different output for every input, whereas the function $h$ has different inputs giving the same output. For example, $h(-3) = 3$ and $h(3) = 3$. The function $f$ has elements in the codomain which are not in the range. For example, 5 is in the codomain $\mathbb{N}$ but 5 is not the square of a natural number, so 5 is not in the range of $f$. However, all the elements in the codomain of $h$ are in its range.

We can formalise these differences and describe two particular sorts of functions:

- functions for which different inputs *always* give different outputs, called **one-one** functions,
- functions whose range is equal to the codomain, called **onto** functions.

Note that the first statement is *not* saying that one input gives two different outputs. A mapping which did this would not be a function.

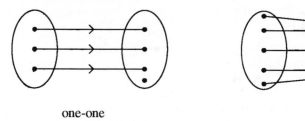

one-one                                                    onto

**Fig 3.12**

Another name for one-one is **injective**. Another name for onto is **surjective**. A function which is both one-one and onto is also called **bijective**.

## ● *Example 25*

From the discussion above it seems that the function

$$f : \mathbb{N} \to \mathbb{N} \text{ where } f(x) = x^2$$

is one-one, but the superficially similar function

$$g : \mathbb{Z} \to \mathbb{Z} \text{ where } g(x) = x^2$$

is not one-one. We shall revisit this example in Example 28.

More formally, we can say that a function $f : A \to B$ is **one-one** (or **injective**) if for all $r, s \in A$ $f(r) = f(s)$ implies that $r = s$.

## ● *Example 26*

Is the following function one-one?

$$f : \mathbb{Z} \to \mathbb{Z} \text{ where } f(x) = x^2 + x + 1$$

SOLUTION
For this function, $f(0) = 1$ and $f(-1) = 1$. Since two different inputs give the same output, $f$ is not one-one (as always, a single counterexample is sufficient).

## ● *Example 27*

Is the following function one-one?

$$f : \mathbb{Z} \to \mathbb{Z} \text{ where } f(x) = 2x - 1$$

SOLUTION
In this example, you might try to find a counterexample, but a little thought (possibly helped by a $y$ against $x$ graph) might convince you that the function is one-one. How can we prove it? Because the number of possible values for the input is infinite, we cannot simply take every single input and show that it gives a different output. We need to use some algebra.

Assume there are two inputs which give the same output, that is let $a, b \in \mathbb{Z}$ such that $f(a) = f(b)$. Then

$$f(a) = f(b) \Rightarrow 2a - 1 = 2b - 1$$
$$\Rightarrow 2a = 2b$$
$$\Rightarrow a = b$$

Hence, $f(a) = f(b) \Rightarrow a = b$. Thus, the only way for two inputs to give the same output is if the two inputs were the same. This proves that $f$ is one-one.

## Example 28

Example 25 revisited. Prove that

$$f: \mathbb{N} \rightarrow \mathbb{N} \text{ where } f(x) = x^2$$

is one-one, but the superficially similar function

$$g: \mathbb{Z} \rightarrow \mathbb{Z} \text{ where } g(x) = x^2$$

is not one-one.

SOLUTION
Let $a, b \in \mathbb{N}$ such that $f(a) = f(b)$. Then

$$f(a) = f(b) \Rightarrow a^2 = b^2$$
$$\Rightarrow a = b \text{ since } a \text{ and } b \text{ are both natural numbers.}$$

Hence, $f(a) = f(b) \Rightarrow a = b$. This proves that $f$ is one-one.
   The function $g$ is not one-one since $g(-2) = g(2)$.

## Example 29

Is the following function one-one?

$$f: \mathbb{N} \rightarrow \mathbb{N} \text{ where } f(n) = \begin{cases} 2n \text{ if } n \text{ is even} \\ n \text{ if } n \text{ is odd} \end{cases}$$

SOLUTION
Let $a, b \in \mathbb{N}$ be such that $f(a) = f(b)$. We have to consider three separate cases: both $a$ and $b$ odd, both even, or one of each.

(i)     $a, b$ both odd:

$$f(a) = f(b) \Rightarrow a = b$$

(ii)    $a$ even and $b$ odd:

$$f(a) = f(b) \Rightarrow 2a = b$$

   This case is impossible since $b$ is odd and therefore can never equal $2a$.

(iii)   $a, b$ both even:

$$f(a) = f(b) \Rightarrow 2a = 2b, \text{ that is } a = b$$

Hence, if $f(a) = f(b)$ then $a = b$, which proves that $f$ is one-one.

Now let us look at some more examples of functions and decide if they are onto or not.

### • *Example 30*

From the discussion at the start of this section it seems that the function

$f: \mathbb{N} \to \mathbb{N}$ where $f(x) = x^2$

is not onto, but the function

$h: \mathbb{Z} \to \{\mathbb{N} \cup 0\}$ where $h(x) = |x|$

is onto. We shall prove this in Example 34.

More formally, we can say that a function $f: A \to B$ is **onto** (or **surjective**) if for each $b \in B$ there is at least one $a \in A$ such that $f(a) = b$.

### • *Example 31*

Let $f: \mathbb{Z} \to \mathbb{Z}$ where $f(x) = x - 1$. Let $r$ be any element of $\mathbb{Z}$ (the codomain). Is there an $x$ in $\mathbb{Z}$ (the domain) so that $f(x) = r$?

SOLUTION
Since $x - 1 = r$, then $x = r + 1$ which is in $\mathbb{Z}$. Thus, $f$ is onto.

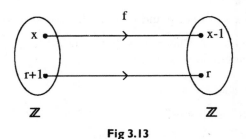

**Fig 3.13**

### • *Example 32*

Let $f: \mathbb{N} \to \mathbb{N}$ where $f(x) = x + 1$. Let $r$ be any element of $\mathbb{N}$. Is there an $x$ in $\mathbb{N}$ such that $f(x) = r$?

SOLUTION
Suppose $r = 1$. Since $x = r - 1$, $x$ would have to be 0. But we require $x$ to be in $\mathbb{N}$. Hence, there is no value of $x \in \mathbb{N}$ such that $f(x) = 1$. Hence $f$ is not onto.

### • *Example 33*

Let $A = \{0, -1, 1\}$ and $B = \{0, 1\}$. Let $f: A \to B$ where $f(a) = |a|$. Is $f$ onto?

SOLUTION
The elements of $B$ are 0 and 1. Hence, $f$ is onto if we can find $x, y \in A$ such that $f(x) = 0$ and $f(y) = 1$. Now, $f(0) = |0| = 0$ and $f(-1) = |-1| = 1$. Hence, for each $b \in B$ there exists an $a \in A$ such that $f(a) = b$. Hence, $f$ is onto.

## ● *Example 34*

Example 30 revisited. Prove that the function

$$f: \mathbb{N} \to \mathbb{N} \text{ where } f(x) = x^2$$

is not onto, but that the function

$$h: \mathbb{Z} \to \{\mathbb{N} \cup 0\} \text{ where } h(x) = |x|$$

is onto.

SOLUTION
The function $f$ is not onto since $3 \in \mathbb{N}$ but there is no $x \in \mathbb{N}$ for which $f(x) = 3$.
The function $h$ is onto since if $x \in \{\mathbb{N} \cup 0\}$ then $x \in \mathbb{Z}$ and $h(x) = |x| = x$.

### EXERCISES ON 3.6

1.  Let $A = \{a, b\}, B = \{c, d\}$. List all the functions $f: A \to B$. Indicate which of the functions are one-one and which are onto.
2.  Give an example of a function $f: \mathbb{N} \to \mathbb{N}$ which is:
    (a)  one-one and onto
    (b)  one-one but not onto
    (c)  not one-one but onto
    (d)  neither one-one nor onto
3.  Let $f: \mathbb{N} \to \mathbb{N}$ where

    $$f(n) = \begin{cases} 2n \text{ if } n \text{ is odd} \\ n \text{ if } n \text{ is even} \end{cases}$$

    Is $f$ one-one? Is $f$ onto?
4.  Let $A = \{0, 1, 2, 3, 4, 5\}$ and $f: A \to A$ where $f(a) = 3a \pmod 6$. Draw a diagram of the function. Is $f$ one-one? Is $f$ onto?
5.  Let $A = \{0, 1, 2, 3, 4, 5, 6\}$ and $f: A \to A$ where $f(a) = 3a \pmod 7$. Draw a diagram of the function. Is $f$ one-one? Is $f$ onto?

## 3.7 Composition of functions

When we want to perform a complicated calculation, we usually break the calculation down into simpler steps. Sometimes we do this almost without thinking.

## ● *Example 35*

Suppose you have a square sheet of paper which measures 10 cm by 10 cm. If you first cut a 1 cm strip off the top edge of the sheet, then cut a 1 cm strip off the left-hand side of the sheet, what is the area of the remaining sheet?

   Since the resulting sheet measures 9 cm by 9 cm, its area is 81cm$^2$.

   What if the original sheet measured $x$ cm by $x$ cm? Then the area of the remaining sheet would have been $(x - 1)^2$ cm$^2$. This formula is the result of two distinct calculations:

● subtract 1 from the length of each side,
● square the result.

We can express this as a *composition* of two functions:

- the 'minus 1' function: $f(x) = x - 1$
- the 'square' function: $g(y) = y^2$

If we let $y$ be the result of applying $f$ to the original side-length, $x$, then $y = x - 1$ and $g(y) = y^2 = (x - 1)^2$, which is the correct result.

In Example 35, we used the function $f$ to convert an input to an output, then used the output from $f$ as the input to another function $g$ to give us our final answer. Since our definition of a function places no restrictions on the domain $A$ or codomain $B$, we can *always* do this: given any function $f: A \rightarrow B$, if $R \subseteq B$ is the range of $f$, we can define another function $g : R \rightarrow C$, and the combined effect of $f$ followed by $g$ will be to map an element of $A$ to an element of $C$. In other words, the combined effect of $f$ and $g$ is to define a third function $h : A \rightarrow C$.

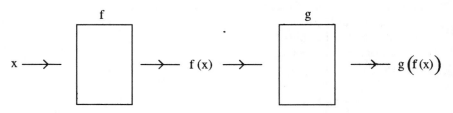

**Fig 3.14**

### Example 36

Let

$$f: \mathbb{Z} \rightarrow \mathbb{Z} \text{ where } f(x) = x - 4$$
$$g : \mathbb{Z} \rightarrow \mathbb{Z} \text{ where } g(x) = x^2 + 1.$$

We can see that $f$ sends 1 to $-3$ and then $g$ sends $-3$ to 10. This can also be written as $g(f(1)) = g(-3) = 10$. So for a general input $x$ we have $g(f(x)) = (x - 4)^2 + 1$.

We read $g(f(x))$ as '$g$ of $f$ of $x$'. Notice that you evaluate the expression *inside out*—you start with $x$, then apply $f$, then apply $g$.

### Example 37

For the same functions as in Example 36 what is $f(g(1))$ and what is $f(g(x))$?

SOLUTION

$$f(g(1)) = f(2) = -2 \text{ and } f(g(x)) = (x^2 + 1) - 4.$$

The order in which we combine two functions matters.

### Example 38

Again for the functions $f$ and $g$ in Example 36, the composition $f(g(x))$ is the function

$$f(g(x)) = (x^2 + 1) - 4 = x^2 - 3$$

Combining $f$ and $g$ in the opposite order gives the composition $g(f(x))$. What function is this? First, $f$ converts $x$ to $x - 4$. Next, $g$ squares $x - 4$ and adds 1. So,

$$g(f(x)) = (x - 4)^2 + 1 = x^2 - 8x + 17$$

Suppose $f : A \to B$ and $g : B \to C$. We define the **composition**, $gf$, to be the function

$$gf : A \to C \text{ where } gf(x) = g(f(x))$$

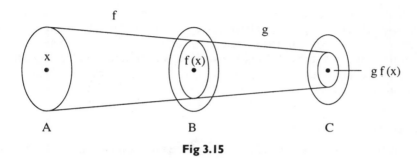

**Fig 3.15**

### Example 39

Let

$$f : \mathbb{Z} \to \mathbb{Z} \text{ where } f(x) = 3x$$
$$g : \mathbb{Z} \to \mathbb{Z} \text{ where } g(x) = x^2$$

What are $gf(2)$ and $fg(2)$? Are there any values of $x$ for which $gf(x) = fg(x)$?

SOLUTION
$gf(2) = g(6) = 36$, $fg(2) = f(4) = 12$. If $gf(x) = fg(x)$ then $(3x)^2 = 3(x)^2$ which is only true for $x = 0$.

### EXERCISES ON 3.7

1. Let $A = \{1, 2, 3\}$. The function $f : A \to A$ has $f(1) = 3$, $f(2) = 1$, $f(3) = 2$. What are $ff$, $fff$ and $ffff$? (Give the values of each function at 1, 2 and 3)
2. Let $f : \mathbb{R} \to \mathbb{R}$ where $f(x) = x^3$ and $g : \mathbb{R} \to \mathbb{R}$ where $g(x) = x - 3$. Give expressions for $ff$, $gg$, $fg$ and $gf$.
3. Let $f : \mathbb{R} \to \mathbb{R}$ where $f(x) = x^2$ and $g : \mathbb{R} \to \mathbb{R}$ where

$$g(x) = \begin{cases} x - 3 & : \quad x < 0 \\ 2x & : \quad x \geq 0 \end{cases}$$

Give expressions for $ff$, $gg$, $fg$ and $gf$.

4. Let $f : A \to B$ and $g : B \to C$. Prove each of the following:
   (a) If $gf$ is one-to-one then $f$ is one-to-one.
   (b) If $f$ and $g$ are both one-to-one then $gf$ is one-to-one.
   (c) If $f$ and $g$ are both onto then $gf$ is onto.

5. Let $f : \mathbb{R} \to \mathbb{R}$ where $f(x) = x^2$ and $g : \mathbb{R} \to \mathbb{R}$ where $g(x) = 3x - 1$. Find the set of values of $x$ for which $fg(x) = gf(x)$.

6. Let $f : A \to B$ and $g : B \to C$. Give a counterexample to each of the following statements:
   (a) If $f$ is one-one then $fg$ is one-one.
   (b) If $fg$ is onto then $g$ is onto.

## 3.8 The inverse of a function

We have seen that a function is a rule for converting one number $x$ into another number $y$, which we express algebraically by writing $y = f(x)$. What if you want to use the rule in the other direction, that is to convert $y$ to $x$ rather than $x$ to $y$? Can we do this for any function? Functions which we can undo in this way are called *invertible*. To undo a function $f$, we need to be able to start with any $y$ in the codomain of $f$ and deduce a unique $x$ in the domain of $f$ such that $f(x) = y$.

### ◉ Example 40

The function $h : \mathbb{Z} \to \mathbb{Z}$ where $h(x) = x - 3$ is easy to undo. If $h(x) = 5$ then $x = 8$ and if $h(x) = 0$ then $x = 3$. If $h(x) = y$ then we can find $x$ since $x = y + 3$.

### ◉ Example 41

Try to undo the functions

$$f : \mathbb{Z} \to \mathbb{Z} \text{ where } f(x) = 3x$$

and

$$g : \mathbb{Z} \to \mathbb{Z} \text{ where } g(x) = x^2$$

What is $x$ if $f(x) = 4$ and what is $x$ if $g(x) = 4$?

SOLUTION
There is no $x \in \mathbb{Z}$ such that $f(x) = 4$. Hence the function $f$ is not onto. There are two possible values of $x$ such that $g(x) = 4$, namely $x = -2$ and $x = +2$. Hence the function $g$ is not one-one.

This example suggests that for a function to be invertible we need it to be both one-one and onto. Before proving this, we need a formal definition of the inverse.

Let $f : A \to B$. The **inverse** of $f$, if it exists, is the function $g : B \to A$ such that, for all $a \in A$ and all $b \in B$, if $f(a) = b$ then $g(b) = a$.

The inverse of $f$ is usually written $f^{-1}$. A function $f$ for which $f^{-1}$ exists is called **invertible**.

## Example 42

Let $f: \mathbf{Z} \to \mathbf{Z}$ where $f(x) = x^2 - 1$. Is $f$ invertible?

SOLUTION

$f(x) = x^2 - 1$ has $f(1) = 0$ and $f(-1) = 0$. So $f^{-1}$ would have to send 0 to 1 and to $-1$. But functions are not allowed to do that. Hence, $f$ is not invertible.

## Example 43

Let $f: \mathbb{N} \to \mathbb{N}$ where $f(x) = x + 3$. Is $f$ invertible?

SOLUTION

An inverse function would have to take any $x \in \mathbb{N}$ back to something. So, in particular, it would have to take 1 back to something. But $f(x) \neq 1$ for any $x \in \mathbb{N}$. Hence, $f$ is not invertible.

## Example 44

Let $f: \mathbf{Z} \to \mathbf{Z}$ where $f(x) = x + 4$. Is $f$ invertible?

SOLUTION

$f^{-1}: \mathbf{Z} \to \mathbf{Z}$ where $f^{-1}(x) = x - 4$. Thus, $f$ is invertible.

Notice that the solution to Example 44 is legitimate because if $x \in \mathbf{Z}$ then $(x - 4) \in \mathbf{Z}$ also. However, in Example 43, we could not define $f^{-1}(x) = x - 3$ because if $x \in \mathbb{N}$ it does not follow that $(x - 3) \in \mathbb{N}$.

We now prove that a function is invertible if and only if it is both one-one and onto, i.e. it is bijective.

## • Theorem I ———————————

Let $f: A \to B$ be a function. The function $f$ is invertible if and only if $f$ is both one-one and onto.

PROOF

Recall that for an 'if and only if' statement to be true, we need to prove two things. Firstly, suppose that $f$ is invertible. We will prove that $f$ is one-one and onto.

Since $f$ is invertible, there is a function $g: B \to A$ such that if $f(a) = b$ then $g(b) = a$. Let $r$ and $s$ be any elements of $A$, and suppose that $f(r) = f(s)$. We shall show that $r = s$. Let $f(r) = f(s) = t$. Then $g(t) = r$ and $g(t) = s$. But $g$ is a function, so $r = s$. Hence, $f$ is one-one. Now, let $b$ be any element of $B$ and define $a \in A$ by $a = g(b)$. Because $g$ is the inverse of $f$, it follows that $f(a) = b$. Hence, $f$ is onto.

We have now proved that if $f$ is invertible, then it is both one-one and onto.

Secondly, suppose that $f$ is one-one and onto. We will prove that $f$ is invertible. Let $b$ be any element of $B$. Since $f$ is onto, there is an $a \in A$ such that $f(a) = b$. Also, since $f$ is one-one there can be no other such $a$. Hence, for each $b$ in $B$ there is a unique $a$ in $A$ with $f(a) = b$. We can therefore define a function $g: B \to A$ by the rule that $g(b)$ is the unique $a$ for which $f(a) = b$. Then, $g$ is the inverse of $f$, so $f$ is invertible.

We have now proved that if $f$ is one-one and onto, then it is invertible.

### EXERCISES ON 3.8

1. Find the inverse function (if it is defined) of the following functions:
   (a) $f: \mathbb{R} \to \mathbb{R}$ where $f(x) = 3x + 2$
   (b) $f: \mathbb{R}^+ \to \mathbb{R}^+$ where $f(x) = (x-1)/(x+2)$
   (c) $f: \mathbb{Z} \to \mathbb{Z}$ where $f(x) = x^2$
   (d) $f: \mathbb{N} \to \mathbb{N}$ where $f(n) = \begin{cases} 2n \text{ if } n \text{ is even} \\ n \text{ if } n \text{ is odd} \end{cases}$

2. Let $f: \mathbb{R} \to \mathbb{R}$ where $f(x) = x^2$ if $x \geq 1$ and $f(x) = x$ if $x < 1$. Find an expression for the inverse of $f$.

3. If $A$ has $a$ elements and $B$ has $b$ elements, how many functions are there from $A$ to $B$? How many of these functions are invertible?

4. Let $f: \mathbb{R} \to \mathbb{R}$ where $f(x) = ax + b$ and $a, b$ are real numbers with $a \neq 0$. Show that $f$ is invertible.

5. Let $f: A \to B$ and $g: B \to C$ both be invertible. Show that $gf$ is invertible and find $(gf)^{-1}$ in terms of $f^{-1}$ and $g^{-1}$.

## 3.9 The pigeonhole principle

In this section we shall show how a very simple result can solve a wide variety of apparently unrelated problems. Consider the following very simple examples:

### Example 45

I have a drawer containing 10 pairs of socks, but all the pairs have been separated. The house is in total darkness. How many socks do I need to take to be sure that I have at least one matching pair?

SOLUTION
If I take 10 socks or fewer, it is possible that they will all be different. But if I take any 11 socks, these must include at least one pair.

### Example 46

There are 13 students sharing a kitchen. Whenever one of them has a birthday, he bakes a cake. Show that there is at least one month when there will be more than one cake.

SOLUTION
There are 13 students, but only 12 months. So it is not possible for all their birthdays to be in different months, and there must be at least one month with at least two cakes.

We can formalise the argument used here as follows.
   **The pigeonhole principle.**   If there are more than $n$ objects to be distributed in $n$ pigeonholes then there is at least one pigeonhole with two or more objects in it.

We can think of the pigeonhole principle in terms of functions. Let $f: A \rightarrow B$, with $|B| = n$ and $|A| = n + 1$. Then there must exist two distinct elements $a$ and $a'$ in $A$ such that $f(a) = f(a')$.

## Example 47

Amongst any eight people there are at least two who were born on the same day of the week.

### SOLUTION
We can think of the eight people as the objects and the seven days of the week as the pigeonholes. We assign the people to the pigeonhole of the day of the week they were born. Since there are more objects than pigeonholes, there must be at least two objects assigned to the same pigeonhole and hence there are two people who were born on the same day of the week.

The pigeonhole principle is so simple that you might expect it could only help you to solve equally simple problems. In fact, it provides *simple* solutions to some quite subtle problems, as we shall now see.

## Example 48

How many people do you need in a school to guarantee that there are two people who have the same initials (first and last names only)?

### SOLUTION
There are 26 letters for each of the first and last names, hence $26^2 = 676$ possible sets of initials. So, by the pigeonhole principle, provided there are more than 676 people, there must be at least two people with the same initials.

In Example 48, the possible sets of initials represent the $n = 676$ pigeonholes, and the people in the school represent the objects. This is still a reasonably simple example which you could probably have worked out for yourself without using the pigeonhole principle explicitly.

## Example 49

Show that in any room of people who have been doing some handshaking there will always be at least two people who have shaken hands the same number of times.

### SOLUTION
This time we label the pigeonholes with the different numbers of hands shaken and put the people (objects) into their correct pigeonhole. Suppose there are $n$ people, then, since people only shake hands with each person at most once, the labels on the pigeonholes will go from 0 to $n - 1$. That is, we have $n$ holes and $n$ people. However, can there be a person in the hole labelled 0 and a person in the hole labelled $n - 1$? It is not possible for the 0th and the $(n - 1)$th holes both to be occupied, because if one person has shaken hands with nobody, then there cannot be any one person who has shaken hands with every other person. Thus, we have at most $n - 1$ holes occupied at any one time. Hence, by the pigeonhole principle at

least one of the holes has two occupants, which shows that there are at least two people who have shaken hands the same number of times.

The next example is a bit more abstract. As always, the trick in solving the problem is to decide what are the objects and what are the pigeonholes.

## ● *Example 50*

Show that however you choose four distinct numbers from 1, 2, 3, 4, 5, 6, there will always be two that sum to 7.

### SOLUTION
The pairs of numbers that sum to 7 are $1 + 6$, $2 + 5$ and $3 + 4$. So we need to show that any subset always has at least one of these pairs. Let the three pigeonholes be the pairs $\{1,6\}$, $\{2,5\}$ and $\{3,4\}$, and the objects the four numbers chosen. Put each number into its own pigeonhole, i.e. a 1 or a 6 goes into the first hole, a 2 or a 5 into the second hole and a 3 or a 4 into the third hole. By the pigeonhole principle, there must be two objects in one of the holes. Since all four numbers are distinct, these two must be different, and must therefore be one of the pairs that sum to 7.

## ● *Example 51*

A hall of residence has 20 rooms and a rota of 25 cleaners. At any one time there will be 20 cleaners on duty. How many room keys do we need to distribute to the 25 cleaners so that whichever 20 are on duty on a particular day, they can always clean every room?

### SOLUTION
Remember that we are looking for the *smallest* number of keys which will do the trick. Obviously, if we gave each cleaner keys to all 20 rooms, there would be no problem. This would need 500 keys. So we know that if $N$ is the solution, then $N \leq 500$. Can we do better? Another way is to give 20 of the cleaners one different key each, so that between them they have access to all 20 rooms, and to give the other 5 cleaners all 20 keys each. Why does this work? Think of any particular room, say room 1. There are 6 keys to room 1, and they are held by 6 different cleaners. Since only 5 cleaners are off-duty on any day, the off-duty cleaners between them have at most 5 keys to room 1, so the on-duty cleaners have at least one key to room 1. If we put this argument in terms of the pigeonhole principle, we are saying that if we try to put the 6 objects (keys to room 1) into 5 pigeonholes (cleaners off duty), then at least one pigeonhole (cleaner) must have at least 2 objects (keys). But this is a contradiction, since we know that no cleaner has two keys to the same room. So the 6 keys to room 1 cannot all be held by the 5 off-duty cleaners, i.e. there must be at least one on-duty cleaner with a key to room 1. The same argument applies to all 20 rooms. So we now know that $N \leq 120$.

Now, 120 is a lot less than 500, but it still seems rather excessive. However, the important part of the argument to show that 120 keys would do the trick was that there were 6 keys to every room ($120 = 20 \times 6$). Now, if we only give out 119 keys then, however we distribute them amongst the cleaners, there will be at least one room with at most 5 keys available for it, and therefore at most 5 cleaners with a key to that room. If all of the cleaners who hold a key to that room are off duty, then the 20 cleaners on duty that day will be unable to open that room. So the solution is $N = 120$.

A simple extension of the pigeonhole principle is to consider a function $f : A \rightarrow B$, where $|B| = n$ and $|A| = kn + 1$, where $k$ is a positive integer. Then, it must be possible to find at least $k + 1$ distinct elements of $A$, say $a_i : i = 1, \ldots, k+1$, such that $f(a_1) = f(a_2) = \ldots = f(a_{k+1})$. Put less formally, if you assign $kn + 1$ objects to $n$ pigeonholes, there must be at least one pigeonhole which contains at least $k + 1$ objects.

## Example 52

Among six people there are either three who all know each other or three who are complete strangers.

SOLUTION
Consider one of the people, say $X$. Of the other five people we can put them in either of two pigeonholes depending on whether they know $X$ or not. In terms of the extended pigeonhole principle, we are thinking of $n = 2$ and $k = 2$. So there must be either $k + 1 = 3$ people who know $X$, or $k + 1 = 3$ people who do not know $X$. Suppose there are three people $A$, $B$ and $C$ who all know $X$. Then if any two of $A$, $B$ and $C$ know each other, say $B$ and $C$, then $X$, $B$ and $C$ all know each other. On the other hand, if no two of $A$, $B$ and $C$ know each other then $A$, $B$ and $C$ form a set of three complete strangers.

A similar argument will apply if we suppose that there are three people who do not know $X$.

## Example 53

There are two people in Britain with the same three initials.

SOLUTION
There are about 55 million people in Britain and the maximum number of different combinations of three initials is $26 \times 26 \times 26 = 17576$. In fact there will be at least 3000 people with the same initials!

## EXERCISES ON 3.9

1.  How many students, each of whom comes from one of the 55 counties of Britain, must there be in the class to be certain there are at least two from the same county?
2.  True or false? If there are more than $m$ objects and there are $m$ pigeonholes:
    (a) there will be at least one pigeonhole with no objects,
    (b) there will be at least one pigeonhole with at least two objects,
    (c) there will be at least one empty pigeonhole,
    (d) there will be at least two pigeonholes with the same number of objects.
3.  Prove that amongst any $n + 1$ arbitrarily chosen integers there will always be two whose difference is divisible by $n$.
4.  How many people do you need in a room to be sure that two will have a birthday in the same month?

5. Suppose that we are given a set of 5 positive integers, where the sum is not greater than 30. Show that we can find two disjoint non-empty subsets of our set whose elements have the same sum.

6. Show that however you choose five points inside an equilateral triangle of side 1, there will be at least two points whose distance apart is less than $\frac{1}{2}$.

7. I make my own dartboard with the numbers 1 to 20 arranged in an arbitrary order in a circle. Show that
   (a) there will always be three numbers in consecutive positions around the board whose sum is at least 32,
   (b) there will always be four numbers in consecutive positions around the board whose sum is at least 42.

8. Show that any sequence of 10 different numbers will contain a subsequence of four numbers which are in either increasing or decreasing order. Find such a sequence that does not have an increasing or a decreasing subsequence of length five.

9. Prove that exactly one of any $n$ consecutive integers is divisible by $n$.

# 3.10 Projects

## Puzzle of the four interconnecting rings

Can you connect three rings so that all three are connected, but if any one ring is broken then all three rings will be separated?

Call the rings $A$, $B$ and $C$. Since no two rings are connected, we can say whether $A$ is above or below $B$. Define a relation **R** that means 'is above'. For the three rings shown in Fig 3.16, we can draw a diagram of the relation (Fig 3.17) to show how the rings $A$, $B$ and $C$ are related.

Can you connect four rings so that all four are connected, but if any one ring is broken then all four rings will be separated? Try to draw a diagram of the relation for the four rings.

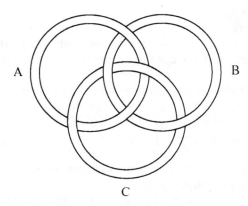

**Fig. 3.16**

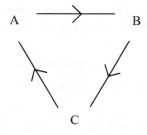

**Fig 3.17**

### Non-transitive dice

Suppose that you have three dice. Die $A$ has the numbers 6, 6, 2, 2, 2, 2 on its six sides. Die $B$ has the numbers 5, 5, 5, 5, 1, 1. Die $C$ has the numbers 4, 4, 4, 3, 3, 3. You challenge a friend to the following game. Your friend first chooses any die, after which you choose one of the two remaining dice. Each of you then throws your die and the person with the higher score wins. Is the game fair?

After a little experimentation you should be able to see that if you played this game for a long time, die $A$ would beat die $B$ more than half the time and die $B$ would beat die $C$ more than half the time, but die $C$ would beat die $A$ more than half the time. This surprising result is a consequence of the fact that the relation 'beats' on the set of three dice is non-transitive.

For the dice $A$ and $B$, draw up a table with six rows and six columns, showing which die wins for each possible combination of results. Count the number of wins for $A$ and the number for $B$. Do the same for dice $A$ and $C$, and for dice $B$ and $C$. What should be your strategy for ensuring that you have an unfair advantage?

Notice that no strategy can *guarantee* that you win the game. This is because the result of each game involves an element of chance, i.e. even if $A$ beats $B$ in the long run, $B$ still has a chance (quite a good chance, in fact) of beating $A$ in any one game. The formal mathematics of chance is called *probability theory*, and is the subject of Chapter 5. After you have read that chapter, you should be able to answer the following questions. For your chosen strategy, what is the probability that you will win any one game? What is the probability that you will win more than half of a sequence of $n$ games?

## *Summary*

Relations:
A relation **R** on a set $S$ is said to be **symmetric** if

$a\mathbf{R}b \Rightarrow b\mathbf{R}a$ for all elements $a, b$ in $S$

A relation **R** on a set $S$ is said to be **transitive** if

$a\mathbf{R}b$ and $b\mathbf{R}c \Rightarrow a\mathbf{R}c$ for all elements $a, b, c$ in $S$

A relation **R** on a set $S$ is said to be **reflexive** if

$a\mathbf{R}a$ for all elements $a$ in $S$

Any relation on a given set $S$ which is symmetric, transitive and reflexive is called an **equivalence relation**. An equivalence relation always partitions $S$ into classes, called **equivalence classes**.

A relation **R** on a set $S$ is said to be **antisymmetric** if

$$a\mathbf{R}b \text{ and } b\mathbf{R}a \Rightarrow a = b \text{ for all elements } a, b \text{ in } S$$

Relations which are reflexive, antisymmetric and transitive on a set are called **partial orders**. The set is then called a **partially ordered set** or **poset**.

Functions:
A **function** $f$ from a set $A$ to a set $B$ is a rule which associates with each member of $A$ a unique member of $B$. We write $f : A \rightarrow B$. We call $A$ the **domain** of the function and $B$ the **codomain**. The subset of $B$ which is actually used by the function is called the **range**, i.e. $\{f(x) \in B : x \in A\}$.

A function $f : A \rightarrow B$ is **one-one** (or **injective**) if, for all $r, s \in A, f(r) = f(s)$ implies that $r = s$.

A function $f : A \rightarrow B$ is **onto** (or **surjective**) if for each $b \in B$ there is at least one $a \in A$ such that $f(a) = b$.

Let $f : A \rightarrow B$ and $g : B \rightarrow C$. The **composition**, $gf$, is the function

$$gf : A \rightarrow C \text{ where } gf(x) = g(f(x))$$

Remember, first do $f$ then do $g$!
Let $f : A \rightarrow B$. The **inverse** of $f$, if it exists, is the function, $g : B \rightarrow A$, such that for all $a \in A$ and all $b \in B$, if $f(a) = b$ then $g(b) = a$. The inverse of $f$ is written $f^{-1}$. A function $f$ for which $f^{-1}$ exists is called **invertible.**

The **pigeonhole principle** is that if there are more than $n$ objects to be distributed in $n$ pigeonholes then there is at least one pigeonhole with two or more objects in it.

# 4 • Combinatorics

What do the following problems all have in common?

- Six friends eat dinner together every night around a circular table. They want to know how often they can have dinner without sitting in the same arrangement.
- A friend of mine known only as Mr X has been collecting Yale keys to help him in his work. He reasoned that there must be a finite number and asked me how many keys he needs so that he has one that will open any door.
- The computer manager at the University of Lancaster needs passwords for 6000 students. The passwords can use any of the letters of the alphabet. What is the minimum length of the passwords so that she can allocate everyone a different password?

All these problems are very simple to state, but to solve them we need some careful counting techniques. Combinatorics is the branch of mathematics which deals with counting problems. In this chapter we introduce general methods which will work for the three problems above but, more importantly, will also give us the tools we need to solve whole classes of counting problems. Many of these problems concern the number of different *selections* which can be made from a given set of objects, and we show you how to deal with four different kinds of selection problem, according to whether repetition is allowed, and whether the order of selection matters. We then discuss the *binomial theorem* and its extension to *multinomials*.

## 4.1 History

Combinatorics has its origins in ancient Chinese and Hindu culture. For example, the medical book of the Hindu, *Susruta*, from around the 6th century BC contains a discussion of the different flavours that can be formed by taking some number of six basic tastes: sweet, acid, saline, pungent, bitter and astringent. *Susruta* says that taking four of the six gives 15 different flavours. There are several of today's well-known counting puzzles whose origins can be traced back many centuries. One example is the children's rhyme, 'As I was going to St Ives', which can be traced back to Fibonacci (1202) and even further back to the Rhind Papyrus (about 1650 BC). You can read more about this in the article 'The Roots of Combinatorics' by N. L. Biggs, in *Historia Mathematica*, volume 6, pages 109–136.

## 4.2 Sum and product

How can we unravel a complex counting problem? We begin with some simple examples where we can easily divide up our problem into independent parts.

## Example 1

A baker has 5 Wholemeal, 3 Granary, 2 Farmhouse and 3 Sliced loaves left at the end of the day. A customer rushes in as they are about to close. How many loaves can he choose from?

SOLUTION
Since the breads are all different, we can add the quantities of the different loaves. The customer has a choice of 13 loaves.

## Example 2

The mathematics teaching committee requires one student representative from either the first year or the second year or the third year. If there are 50 first year, 30 second year and 10 third year students, how many different representatives are there?

SOLUTION
Since the students are all different we again add to get 90 different possible representatives.

Although the practical contexts, and the numbers used, are clearly different for these two problems, the mathematical process by which we solve the problem is the same in both cases: we find the total number of objects in a set by splitting the set into disjoint subsets and adding the numbers of objects in each subset. We can state this more formally as follows.

**Rule of Sum.** If a set of objects can be divided into $m$ disjoint subsets and the $i$th subset has $n_i$ elements, then we have exactly $n_1 + n_2 + \ldots + n_m$ objects altogether.

## Example 3

The library has 20 books by Dorothy Sayers and 45 books by Agatha Christie. By the rule of sum a reader has a choice of 65 books by either author.

## Example 4

The town council has 20 Labour, 15 Conservative and 5 Independent Councillors. Every week a different councillor chairs the meeting. By the rule of sum there are 40 weeks until a repeat chair is necessary.

## Example 5

I can speak three languages and my friend can speak six languages. If $n$ is the number of languages we can speak between us then $6 \leq n \leq 9$. In this example the two subsets are not necessarily disjoint.

Our next examples introduce another rule.

## Example 6

A mixed pair are to be chosen to represent the local ice skating club. The club has 5 female and 7 male skaters. How many different pairs are there?

SOLUTION
There are 5 ways to choose the woman and for each of those women there are 7
possible men. So the total number of pairs is $5 \times 7 = 35$.

## Example 7

A pharmaceutical company is testing a new drug. The drug has 3 components,
which we shall call A, B and C. Each of the components can occur at different
levels. There are 3 levels of A, 4 levels of B and 2 levels of C. How many different
variations of the drug can they make?

SOLUTION
There are 3 ways to choose the amount of A, and for each level of A there are
4 levels of B so there are 12 A–B combinations. For each of these there are 2
possible levels of C. Hence, there are $4 \times 3 \times 2 = 24$ different varieties of this drug.

## Example 8

The mathematics student committee is composed of three students: one from each
of the first year, the second year and the third year. If there are 50 first years, 30
second years and 10 third years, how many possible different committees are there?

SOLUTION
There are $50 \times 30 \times 10 = 15000$ different ways of choosing the committee.

In each of these examples we are required to choose an object from each of several
different sets. The total number of possibilities can be obtained by multiplying
together the numbers of objects in each set. As with our earlier examples, our
solutions use the same mathematical argument, but in different practical contexts.
We can state the mathematical argument more formally as follows.

**Rule of Product.** If an activity can be composed from $m$ different objects, of which
the $i$th can be chosen in $n_i$ ways, then we have exactly $n_1 \times n_2 \times \ldots \times n_m$ ways of
composing the activity.

## Example 9

The stalls in an outdoor market are identified by a letter and a single number. What
is the maximum number of stalls?

SOLUTION
There are 26 possible letters and 10 possible numbers (including 0). So, by the rule
of product there can be a maximum of $26 \times 10 = 260$ stalls.

## Example 10

The university music group is performing Mozart's piano trio in G, for which a
violinist, a pianist and a cellist are needed. There are 12 violinists, 3 pianists and 4
cellists in the group. How many different trios can be formed?

SOLUTION
By the rule of product we have $12 \times 3 \times 4 = 144$ possible trios.

In more complex counting problems we need to use both the sum and product rules in combination.

## ● *Example 11*

I am allowed to take two pieces of luggage with me on my trip to America. I have 3 suitcases, 4 rucksacks and 2 holdalls. How many ways can I choose two different types of luggage?

SOLUTION

The first step in the solution is to break down a complex problem into several simpler sub-problems. First, notice that we must choose either a suitcase and a rucksack (call this SR), a suitcase and a holdall (SH) or a rucksack and a holdall (RH). By the multiplication rule we can determine the number of SR choices as $3 \times 4$, the number of SH choices as $3 \times 2$ and the number of RH choices as $4 \times 2$. Since all of these choices are different, and there are no other possibilities, we use the rule of sum to deduce that there are a total of $12 + 6 + 8 = 26$ different luggage combinations.

Notice that the key to solving the problem is to organise the calculations so that we count every possible choice once and only once. What would the answer have been if we had also been allowed to choose two suitcases as our two pieces of luggage?

## EXERCISES ON 4.2

1. A student wishes to take just one course from the science faculty. There are 3 Physics, 4 Chemistry and 3 Biology courses on offer. How many different choices can the student make?

2. Another student wishes to take a combination of three courses, one from each of the three science departments in Exercise 1. How many possible combinations are there?

3. The science faculty of Exercise 1 is offering a two-course minor, which involves a course from each of two different departments. How many such minors can they advertise?

4. I like to vary my daily journey from my home to work at the university. I always travel via Lancaster. From my home to Lancaster I can take 3 different routes and from Lancaster to the university there are 4 different routes. How many different return routes are there?

5. A binary word is made up of 0's and 1's. How many binary words are there with length exactly 7?

6. British car number plates are of the form

    C 506 DBC

    where the first letter represents the year of manufacture. What is the maximum number of cars that can be licensed in any one year?

7. A palindrome is a string of digits that reads the same backwards as forwards. How many different palindromes are there with 5 digits? How many with 6?

8. How many ways are there of arranging 5 different books on a shelf?

9. In Morse code, each letter of the alphabet is represented by a sequence of dots and/or dashes. How many different symbols can be represented using sequences of length at most 4?
10. How many numbers between 1 and 1000 have exactly one 7 in them?

# 4.3 Permutations and combinations

You are offered a bag containing three kinds of sweets, Allsorts, Butterscotch and Cherry Drops. In how many ways can you choose 2 sweets? We shall use the initial letters, A, B and C, to denote the three types of sweet. Before we can answer the question, we need to be sure of its precise meaning. Do we allow the same type of sweet to be chosen twice—is AA allowed? Does it matter in what order the sweets are chosen—is AB the same choice as BA? Put more formally, we need to decide whether we allow repetition or not, and whether the order matters or not. This gives us four cases to consider.

- Allow repetition and order matters: AA AB BA BB AC CA CC BC CB.
- Allow repetition and order does not matter: AA AB BB AC CC BC.
- Do not allow repetition and order matters: AB BA AC CA BC CB.
- Do not allow repetition and order does not matter: AB AC BC.

Before we can solve any selection problem we have to decide whether repetition is allowed, and whether the order of the selection is important. In the following examples, try to decide which type of selection is required before reading the solutions.

## Example 12

Six friends go to an ice cream parlour where there are 4 different flavours of ice cream on sale: vanilla, tutti-frutti, chocolate and mint-chip. How many different selections can they make?

SOLUTION

Since the friends do not have to choose different ice cream flavours we allow repetition. The order matters because we can organise the friends by the numbers 1 to 6 with person 1 having the first ice cream chosen, person 2 the second and so on. So this is an ordered selection with repetition, of 6 objects (the ice creams chosen by the six friends) from 4 (the flavours available).

## Example 13

In how many ways can a committee with 10 members select a chairperson, a treasurer and an administrator?

SOLUTION

If we say that the first person chosen is the chairperson, the second the treasurer and the third the administrator then we have an ordered selection. Since we do not want anybody having more than one job we do not allow repetition. Hence this is an ordered selection without repetition, of 3 from 10.

## Example 14

An exam has 7 questions of which the student must answer 4. How many different question selections are there?

SOLUTION

The student certainly does not want to do a question more than once so we do not allow repetition. The order in which the student answers the questions does not matter. So this is an unordered selection without repetition, of 4 objects from 7.

## Example 15

How many different patterns can you get by throwing three identical dice?

SOLUTION

We can think of this as selecting three numbers, each between 1 and 6 inclusive. Since the dice are identical, the order does not matter. Also, we are allowed to repeat a number. So this is an unordered selection with repetition, of 3 from 6.

### Ordered selections with repetition

Six friends go to an ice cream parlour which sells 4 different flavours of ice cream: vanilla, tutti-frutti, chocolate and mint-chip. How many different orders can they get? We saw that this is an example of an ordered selection with repetition. The first person has a choice of 4 different flavours, the second has a choice of 4 different flavours, as does the third and so on up to the last person. So, by the multiplication rule the number of different orders is $4 \times 4 \times 4 \times 4 \times 4 \times 4 = 4^6$.

We can state this result formally as follows:

## • Theorem 1 ——————————————

If repetition is allowed then the number of ordered selections of $k$ objects from a set of $n$ objects is $n^k$.

PROOF

There are $n$ ways to select the first object. After we have selected the first object, there are $n$ ways to select the second object. We continue selecting objects until we have selected all $k$ objects. By the rule of product these selections can be made in $n \times n \times \ldots \times n = n^k$ ways.

## Example 16

How many different five-figure telephone numbers are there?

SOLUTION

There are 10 possible digits, $0, 1, \ldots, 9$ for each of the 5 figures. Thus we have an ordered selection with repetition of 5 objects from 10. There are $10^5$ telephone numbers of five digits.

This solution assumes that numbers with leading zeros are allowed. How many different numbers are there if leading zeros are not allowed? How many different numbers are there if leading zeros are not allowed and the numbers may have less than five digits?

## Example 17

In the Chinese book the I Ching, or Book of Changes, there are various prophesies which depend on a certain combination of two symbols. The two symbols are called Yin, a broken line symbolising earth, and Yang, a solid line symbolising heaven. Each combination of six of these symbols is called a hexagram. For example, the hexagram drawn in Fig 4.1 is the symbol of Peace. How many possible different hexagrams are there?

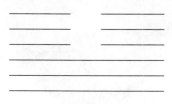

**Fig 4.1**

SOLUTION
There are $2^6 = 64$ different symbols possible, and indeed the I Ching uses all 64.

## Example 18

A locksmith is designing a new key which has 5 different positions, each of which can be cut to 8 different depths. How many different keys can she sell?

SOLUTION
This is an ordered selection with repetition, of 5 objects from 8. There are $8^5$ different keys.

## Ordered selections without repetition

In how many ways can a committee with 10 members select a chairperson, a treasurer and an administrator? Suppose that the first person selected is to be the chairperson, who can therefore be chosen in 10 ways. Suppose the second person selected is the treasurer, who can therefore be chosen in 9 ways. Finally, suppose the last person chosen is the administrator, for whom there are 8 possible people available. So, using the product rule the number of selections is $10 \times 9 \times 8 = 720$.

## ● Theorem 2 ─────────────────────────────

If repetition is not allowed the number of ordered selections of $k$ objects from a set of $n$ objects is $n(n-1)(n-2)\ldots(n-k+2)(n-k+1)$, written $(n)_k$.

PROOF
There are $n$ ways to select the first object, but now there are one fewer objects, $n-1$, from which to choose the second object. We continue selecting elements until the last object, which can be chosen in $n-k+1$ ways. By the product rule, these selections can be made in $n(n-1)(n-2)\ldots(n-k+1)$ ways.

**Fig 4.2** Jakob Bernoulli (David Eugene Smith Collection, Rare Book and Manuscript Library, Columbia University)

This type of selection is known as a **permutation**, a name given by Jakob Bernoulli (1654–1705). Bernoulli was a Swiss mathematician and physicist. His book *Ars Conjectandi*, published eight years after his death, gives the first systematic treatment of combinatorics and its relationship to the theory of probability (see Chapter 5).

We shall often write expressions like $4 \times 3 \times 2 \times 1$, so we give this quantity a special name, 4 factorial, and write it as $4!$.

For positive integers $n$,

$$n! = n(n-1)(n-2)\ldots(3)(2)(1)$$

Also,

$$0! = 1$$

## Example 19

We can use factorial notation to express the value of $(n)_k$ as

$$(n)_k = \frac{n!}{(n-k)!}$$

## Example 20

In a competition to win a holiday in France, the following attractions of the country are listed: Haute Cuisine, Climate, Riviera, EuroDisney, Vineyards, and Friendliness. To win the competition you have to select the most important four of these in order. How many entries would you have to make to ensure that one of your answers was correct?

SOLUTION
This is an ordered selection without repetition of 4 from 6. The number of entries needed is $(6)_4 = 6!/(6-4)! = 360$.

### ● *Example 21*

There is a story told of a German called Leuben, who bet that he could eventually shuffle a pack of cards so that it would appear in a particular order. He actually succeeded, after shuffling for 20 years at 10 hours a day, in 4 246 028 shuffles! This is rather remarkable since the number of different orders for the cards is 52!, which is approximately $8 \times 10^{67}$.

### ● *Example 22*

This example is taken from Bhaskara's *Lilavati* (1150). How many ways are there of arranging the 10 objects in the god Sambhu's 10 hands? The 10 objects are the rope, the elephant's hook, the serpent, the tabor, the skull, the trident, the bedstead, the dagger, the arrow, and the bow.

SOLUTION
The number of rearrangements is $10! = 3\,628\,800$.

### ● *Example 23*

How many ways are there of seating 6 people around a circular dinner table, if rotations of any particular arrangement are considered to be equivalent?

SOLUTION
If we just wanted to know the number of ways of seating 6 people in a line then the solution would be $(6)_6 = 6!$. Call the 6 people A, B, C, D, E, F and consider the permutation BEDCAF, say. Then, EDCAFB, DCAFBE, CAFBED, AFBEDC and FBEDCA all give the same seating plan as BEDCAF. Hence there are $6!/6 = 5!$ ways of seating the 6 people.

### Unordered selections without repetition

An exam has 7 questions of which the student must answer 4. How many different question selections are there? First of all we could count the number of selections assuming that the order mattered. The solution is $7 \times 6 \times 5 \times 4 = 840$. Look at a typical selection, say Q2, Q7, Q3, Q5. There are $4! = 24$ ways of ordering these questions, so we need to divide 840 by 24. This gives just 35 ways of selecting the exam questions. More generally, we have the following result.

### ● *Theorem 3* ———————————————————————

If repetition is not allowed then the number of unordered selections of $k$ objects from a set of $n$ objects is

$$\frac{(n)_k}{k!} = \frac{n(n-1)\ldots(n-k+1)}{k!} = \frac{n!}{(n-k)!k!}$$

written $\binom{n}{k}$.

PROOF
The number of ways of selecting without repetition $k$ ordered objects from $n$ is the same as the number of ways of first selecting without repetition $k$ unordered

objects, then taking each of these selections and ordering them in all possible ways. Thus,

$$(n)_k = \binom{n}{k}k!$$

This type of selection is known as a **combination.** The notation $\binom{n}{k}$ was introduced in the nineteenth century by Andreas von Ettinghausen.

## Example 24

In the card game bridge, a hand has 13 cards. How many different bridge hands are there?

SOLUTION

This is an unordered selection without repetition, of 13 from 52. Hence, there are $\binom{52}{13}$ different bridge hands.

## Example 25

The mathematics students are entering a team of 4 in the university relay race. There are 30 undergraduates and 20 postgraduates. How many teams are possible:

(a)  if any student is eligible?
(b)  if there must be 2 postgraduates and 2 undergraduates?
(c)  if one of the undergraduates and one of the postgraduates are county runners and must both be chosen?

SOLUTION

(a)  There are $\binom{50}{4} = 230\,300$ possible teams.
(b)  The 2 undergraduates can be chosen in $\binom{30}{2} = 435$ ways and the postgraduates in $\binom{20}{2} = 190$ ways. By the product rule there are $435 \times 190 = 82\,650$ possible teams.
(c)  One of the two undergraduates is fixed, the other can be chosen in $\binom{29}{1} = 29$ ways. One of the postgraduates is fixed, the other can be chosen in $\binom{19}{1} = 19$ ways. By the product rule, there are $29 \times 19 = 551$ possible teams.

Notice how the number of possibilities shrinks dramatically as you impose more and more restrictions on the problem.

## Example 26

Here is a puzzle to try out on your friends to see how good is their mathematical intuition. I have a drawer containing three black socks and one white sock. I grab two socks at random. What is the chance of getting a matching pair?

SOLUTION

Altogether there are $\binom{4}{2} = 6$ pairs, of which $\binom{3}{2} = 3$ are a black pair. That is, half the time I would expect to get a matching pair.

## Example 27

This example is taken from the Brhatsamhita of Varahamihira in the 6th century. There are 16 available ingredients for making perfume. How many different perfumes can be made if we use just four ingredients each time?

SOLUTION
There are $\binom{16}{4} = 1820$ different perfumes.

## Unordered selections with repetition

How many different patterns can you get by throwing three identical dice? In order to find a general solution for this problem, we shall show how it is equivalent to another problem which is much easier to solve! A throw of two 3's and a 6 could be illustrated by the following table:

| 1 | 2 | 3 | 4 | 5 | 6 |
|---|---|----|---|---|---|
|   |   | xx |   |   | x |

Illustrate some throws yourself. For example, try 125, 227 or 653.

Every throw has a unique representation on a table of this kind, and every table represents a unique throw. The number of different throws must therefore be the same as the number of ways of placing three x's in the six columns.

One way to represent each row of the table is to write the vertical lines which divide the numbers 1 to 6 as |. With this representation, the sequence of throws 336 becomes ||xx|||x, 125 becomes x|x|||x|, 356 becomes ||x||x|x. Represent the selections 446, 312 and 665 using |'s and x's. What selections do |||xx|x|, |xx|x||| and x|||||xx represent?

Now, each throw of three dice is represented by five |'s and three x's. Therefore, the number of different patterns is the same as the number of ways of writing a sequence of length 8 with exactly five |'s and three x's. As we already know, this is $\binom{8}{3}$. This correspondence works in situations other than dice throwing.

## Example 28

How many ways are there to choose an unordered selection of two letters from $A$, $B$ and $C$ allowing repetition?

SOLUTION
We can represent each possible selection as a table with three columns in which two x's have been placed in the columns. This in turn corresponds to a sequence of two |'s and two x's. For example we can represent a choice of two $B$'s as:

| $A$ | $B$ | $C$ |
|-----|-----|-----|
|     | xx  |     |

or, more compactly, as |xx|. There are $\binom{4}{2}$ such sequences. In other words there are six ways to choose two letters from $A$, $B$ and $C$ allowing repetition.

In general, we can use a similar representation for any unordered selection with repetition allowed.

## ● *Theorem 4*

Given an unordered selection of $k$ objects from a set of $n$ objects, with repetition allowed, we can represent each selection using a sequence of $k$ $x$'s and $(n-1)$ |'s. There are $\binom{n-1+k}{k}$ such sequences.

## ● *Example 29*

In order to get a voucher worth 50p it is necessary to buy three tins of soup. The supermarket stocks chicken, leek, minestrone, tomato and oxtail soup. How many different selections are there?

SOLUTION

This is an unordered selection with repetition, of 3 objects from 5 types. Hence, the number of selections is $\binom{7}{3}$.

## ● *Example 30*

In how many ways can 10 balloons be distributed at a party for 8 children?

SOLUTION

This is an unordered selection with repetition, of 10 objects from 8 types. Hence, the number of selections is $\binom{17}{10}$. If we want to ensure that no child gets left out, we must first give a balloon to each child, then distribute the remaining two balloons, which we can do in $\binom{8+2-1}{2} = \binom{9}{2} = 36$ ways.

## ● *Example 31*

The ancient Egyptians used to play games like dice with the astragalus bone of an animal. When an astragalus is thrown it can land in any one of four ways. How many different possibilities are there if you throw three identical astragali?

SOLUTION

This is an unordered selection with repetition, of 3 objects from 4 types. Hence, the number of selections is $\binom{4+1-3}{3} = \binom{7}{3} = 35$.

We finish this section with a nice example of the use of combinatorics in a problem in physics.

## ● *Example 32*

Within the atom, particles can be in different energy states. Physicists want to know how many different ways $r$ particles can arrange themselves if there are $n$ different energy states.

SOLUTION

The answer depends on how the particles are modelled. There are three well-known models. The Maxwell–Boltzmann model assumes that the particles are all different and that they can be in any energy state. This gives us $n^r$ posssible distributions. This model was found not to be realistic. A second model, the Bose–Einstein model, assumes that the $r$ particles are identical and that each one can be in any energy state. This model gives $\binom{n+r-1}{r}$ ways. Particles such as photons were found to behave in this way. In the third model, the Fermi–Dirac model, the $r$ particles are identical and each energy state can contain at most one particle. This model gives $\binom{n}{r}$ different ways for the particles to be arranged. Particles such as protons or electrons exhibit this behaviour.

## EXERCISES ON 4.3

1. The local hardware store sells paint in 3 different makes, 12 colours, 2 finishes and 4 can sizes.
   (a) How many different cans of paint are on sale?
   (b) If I want blue paint, how many choices can I make?

2. In how many ways can you select a committee of four from a group of seven people?

3. How many four-note bars can you play if you can play any of eight notes and:
   (a) repeating a note is allowed?
   (b) all the notes must be different?

4. Humans get their different characteristics from their chromosomes. We each have 46 chromosomes. Each chromosome is made up of $4 \times 10^9$ nucleotides and each nucleotide is one of the four compounds adenine, cytosine, guanine or thymine. How many genetically different human beings are there?

5. How many ways can six people be seated around a circular table if two of them insist on sitting together? (Note: in this and the next question, all rotations of any particular arrangement are considered to be equivalent.)

6. In how many ways can three couples be seated around a circular table if
   (a) the sexes should alternate?
   (b) the sexes should alternate and no couple should sit together?

7. This question is taken from Bhaskara's *Lilavati* (1150). In a pleasant, spacious and elegant edifice, with eight doors, constructed by a skilful architect for the lord of the land, how many different ways can the doors be opened or closed?

8. A cabaret organiser has to schedule three comedy acts, four musical acts and two acrobatic acts.
   (a) Suppose the acrobats are to perform first, then the musicians and finally the comedians. How many such shows are possible?
   (b) Suppose that the different types of acts must still be grouped together, but that the order of the three types does not matter. Now how many shows are possible?
   (c) If the organiser allows them to perform in any order, how many shows are possible?

9. In the game of Cluedo a suspect is accused of committing a murder in a room with an object. There are six suspects, nine rooms and six murder weapons.
   (a) Initially the players have no clues. How many different accusations are there?
   (b) After a while you realise that the murderer must be either Prof. Plumb or Rev. Green, who used either the lead piping or the rope in either the library or the dining room. Now how many different accusations can you make?

10. A university has 58 male professors and 3 female professors. It is decided that every committee of professors should have at least one female member and that no two committees should have exactly the same members. How many different committees of 12 people can be formed?

11. Seven people are sitting around a table with a plate containing two doughnuts, two macaroons, two cream horns and two eccles cakes. How many ways are there of choosing one cake each?

12. You are buying five New Year cards from a shop which has four types that you like.
    (a) How many possible ways are there to buy the five cards?
    (b) How many possibilities include exactly two of the four types of card?
    (c) How many possibilities include as much variety as possible?

13. The computer manager at the University of Lancaster needs passwords for 6000 students. The passwords can use any of the letters of the alphabet. What is the minimum length of the passwords so that she can allocate everyone a different password?

# 4.4 Pascal's triangle

Blaise Pascal (1623–1662) was a French philosopher and mathematician. Pascal's mathematical talent was discovered early, and by the age of 14 he was discussing

**Fig 4.3** Blaise Pascal (David Eugene Smith Collection, Rare Book and Manuscript Library, Columbia University)

mathematics with some of the best French mathematicians. However, this early start did not continue as in his twenties he decided to devote the rest of his life to religious contemplation. He invented the first working mechanical calculating device in 1645 (see Fig 4.4). The programming language PASCAL is named after him.

One of Pascal's many mathematical discoveries was the following. He noticed that he could obtain the numbers $\binom{n}{k}$ by constructing a triangular array using some

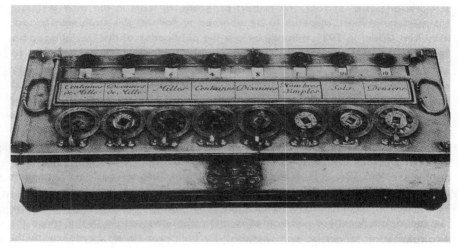

**Fig 4.4** One of Pascal's arithmetic machines, devised by him in 1642

very simple arithmetic. We label the rows of the triangular array $n = 0, n = 1, \ldots$, and the positions within the $n$th row as $k = 0, 1, \ldots, n$. The construction of the array, which is usually called **Pascal's triangle,** then proceeds as follows.

Let the zero row of the triangle be the single entry, 1, and the first row be a pair of entries each equal to 1. This gives the first two rows of Pascal's triangle as

$$
\begin{array}{cc}
 & 1 & \\
1 & & 1
\end{array}
$$

We then calculate the $n$th row of the triangle, which contains $n + 1$ numbers, from the preceding row by the following rules:

1. The first ($k = 0$) and last ($k = n$) entries are both equal to 1.
2. For $1 \le k \le n - 1$, the $k$th entry in the $n$th row is the sum of the $(k - 1)$th and $k$th entries in the $(n - 1)$th row.

The first six rows of Pascal's triangle are shown in the following diagram:

$$
\begin{array}{ccccccccccc}
 & & & & & 1 & & & & & \\
 & & & & 1 & & 1 & & & & \\
 & & & 1 & & 2 & & 1 & & & \\
 & & 1 & & 3 & & 3 & & 1 & & \\
 & 1 & & 4 & & 6 & & 4 & & 1 & \\
1 & & 5 & & 10 & & 10 & & 5 & & 1
\end{array}
$$

Now, suppose you wanted to construct a triangular array of numbers according to the following rules:

1. Rows are labelled $n = 0$, $n = 1, \ldots$ and positions within the $n$th row are labelled $k = 0, 1, \ldots, n$.
2. The $k$th position in the $n$th row receives the number $\binom{n}{k}$.

The first six rows of this array are the following:

$$\binom{0}{0}$$
$$\binom{1}{0} \qquad \binom{1}{1}$$
$$\binom{2}{0} \qquad \binom{2}{1} \qquad \binom{2}{2}$$
$$\binom{3}{0} \qquad \binom{3}{1} \qquad \binom{3}{2} \qquad \binom{3}{3}$$
$$\binom{4}{0} \qquad \binom{4}{1} \qquad \binom{4}{2} \qquad \binom{4}{3} \qquad \binom{4}{4}$$
$$\binom{5}{0} \qquad \binom{5}{1} \qquad \binom{5}{2} \qquad \binom{5}{3} \qquad \binom{5}{4} \qquad \binom{5}{5}$$

If we evaluate the numbers $\binom{n}{k}$, we find that we obtain the same numbers as in the first six rows of Pascal's triangle. For example, the number in the position $n = 5$, $k = 3$ is $\binom{5}{3} = 10$. This is (of course!) no coincidence. In fact, we can prove that Pascal's construction will produce the numbers $\binom{n}{k}$ for every value of $n$ and $k \leq n$. The first part of the proof is to notice that, since $\binom{n}{0} = \binom{n}{n} = 1$ for any value of $n$, the first and last entries in each row of Pascal's triangle give the required result. The more interesting (and slightly harder) part of the proof is to show that Pascal's construction gives the numbers $\binom{n}{k}$ for $1 \leq k \leq n - 1$. This requires us to establish the following identity.

## ● *Example 33*

For any positive integer $n$ and any integer $k$ between 1 and $n - 1$,

$$\binom{n}{k} = \binom{n-1}{k-1} + \binom{n-1}{k}$$

SOLUTION
The algebraic solution is as follows.

$$\binom{n-1}{k-1} + \binom{n-1}{k} = \frac{(n-1)!}{(n-1-(k-1))!(k-1)!} + \frac{(n-1)!}{(n-1-k)!k!}$$

$$= \frac{(n-1)!}{(n-k)!(k-1)!} + \frac{(n-1)!}{(n-1-k)!k!}$$

$$= \frac{(n-1)!k}{(n-k)!k!} + \frac{(n-1)!(n-k)}{(n-k)!k!}$$

$$= \frac{(n-1)!k + (n-1)!n - (n-1)!(k)}{(n-k)!k!}$$

$$= \frac{n!}{(n-k)!k!}$$

$$= \binom{n}{k}$$

This algebraic solution is not very elegant, and you need to be very careful not to make a mistake in the algebra. We shall try a different approach.

The combinatorial solution is as follows. We know that $\binom{n}{k}$ represents the number of ways of choosing $k$ objects from $n$. Now, we can select any one object from the set of $n$ objects and mark it red. Then, when we choose our $k$ objects, either the red one is included or it is not. In the first case, we are choosing $k - 1$

objects from the $n - 1$ objects that are not red, and this can be done in $\binom{n-1}{k-1}$ ways. In the other case, we are choosing $k$ objects from the $n - 1$ objects that are not red, and this can be done in $\binom{n-1}{k}$ ways. So, by the rule of sum, the total number of choices is $\binom{n-1}{k-1} + \binom{n-1}{k}$, which must therefore be equal to $\binom{n}{k}$.

Pascal used his triangle to help him with some problems arising in gambling. Although the triangle is usually ascribed to Pascal, it occurred earlier in the work of the Chinese mathematician Chu Shï-kié (1303) (see Fig 4.5), and in Europe in Petrus Apianus' book *Arithmetic* (1527).

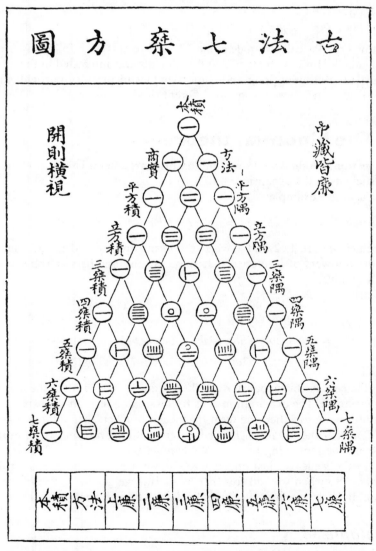

**Fig 4.5** Pascal's arithmetical triangle as depicted in 1303 by Chu Shï-kié

Notice that each row of Pascal's triangle is reversible. This corresponds to the fact that $\binom{n}{k} = \binom{n}{n-k}$.

### EXERCISES ON 4.4

1. Fill in the $n = 7$ and $n = 8$ rows of Pascal's triangle. Notice how much quicker this is than calculating all the numbers $\binom{n}{k}$ directly.
2. Calculate the sum of the numbers in each row of Pascal's triangle. Do you notice a pattern in these numbers?
3. Show both algebraically and combinatorially that

   (a) $\binom{n}{k} = \binom{n}{n-k}$

   (b) $\binom{n}{0} + \binom{n}{1} + \ldots + \binom{n}{n} = 2^n$

   (c) $\binom{n}{0} + \binom{n+1}{1} + \binom{n+2}{2} + \ldots + \binom{n+r}{r} = \binom{n+r+1}{r}$

4. Show how Pascal's triangle could be used to solve the following problem of Pingala, a Hindu scholar of 200 BC. Pingala was interested in the numbers of ways of combining long and short syllables to produce particular rhythms. With three syllables how many different rhythms are there?

## 4.5 The binomial theorem

In this section, we shall see how combinatorial arguments can be used to evaluate the coefficients in the expansion of $(a + b)^n$.

First, a simple example:

$$(a + b)^2 = (a + b)(a + b) = aa + ab + ba + bb = a^2 + 2ab + b^2$$

The coefficients are 1, 2 and 1 when the terms are ordered in the usual way in descending powers of $a$. Notice that these are the numbers in the $n = 2$ row of Pascal's triangle.

Similarly

$$\begin{aligned}
(a + b)^3 &= (a + b)(a + b)(a + b) \\
&= aaa + aab + aba + abb + baa + bab + bba + bbb \\
&= a^3 + 3a^2b + 3ab^2 + b^3
\end{aligned}$$

The coefficients in this expansion are 1, 3, 3 and 1, which form the $n = 3$ row of Pascal's triangle. The explanation is as follows. Each term in the expansion arises from multiplying together one term from each bracket, then adding together like terms. Consider the $a^2b$ term. This comes from taking $a$ from two of the brackets and $b$ from the remaining bracket. We can get the $a$'s in $\binom{3}{2} = 3$ ways, and each choice for the $a$'s leaves just one way to choose the $b$. By the product rule we get $3 \times 1$ ways. So when we add together like terms we get $3a^2b$. The $a^3$ term comes from taking $a$ from each of the three brackets. This can be in $\binom{3}{3} = 1$ way only.

### ● *Example 34*

What is the coefficient of $x^2y^2$ in $(x + y)^4$?

SOLUTION

We need to choose an $x$ from two brackets and a $y$ from the remaining two brackets. The $x$'s can be chosen in $\binom{4}{2} = 6$ ways, and each choice for the $x$'s leaves just one choice for the $y$'s. Hence, we have 6 ways. The coefficient of $x^2y^2$ is therefore 6.

● *Theorem 5* ─────────────────────────────

**The binomial theorem.** If $a$ and $b$ are real numbers and $n$ is a positive integer, then

$$(a+b)^n = \binom{n}{0}a^nb^0 + \binom{n}{1}a^{n-1}b^1 + \binom{n}{2}a^{n-2}b^2 + \ldots + \binom{n}{n}a^0b^n$$

PROOF

When we expand $(a+b)^n$ the terms will be of the form $a^{n-k}b^k$, where $k$ takes the values from 0 to $n$. The coefficient of $a^{n-k}b^k$ comes by choosing $b$ from $k$ of the $n$ brackets in the expansion of $(a+b)^n$, and $a$ from the $n-k$ remaining brackets. We can choose the $b$'s in $\binom{n}{k}$ ways and each choice of the $b$'s leaves just one choice for the $a$'s. Therefore, the number of terms $a^{n-k}b^k$ in the expansion is $\binom{n}{k}$.

The binomial theorem for $n = 2$ was probably known in 1800 BC. It occurs in print in about 300 BC in the work of Euclid. Pascal published a paper in 1653 linking binomial coefficients, combinations and the coefficients of polynomials.

● *Example 35*

Write out the expansion of $(a+b)^6$.

SOLUTION

From the binomial theorem

$$(a+b)^6 = \binom{6}{0}a^6b^0 + \binom{6}{1}a^5b^1 + \binom{6}{2}a^4b^2 + \binom{6}{3}a^3b^3 + \binom{6}{4}a^2b^4$$

$$+ \binom{6}{5}a^1b^5 + \binom{6}{6}a^0b^6$$

$$= a^6 + 6a^5b + 15a^4b^2 + 20a^3b^3 + 15a^2b^4 + 6ab^5 + b^6$$

● *Example 36*

Use the binomial theorem to obtain

$$\binom{6}{0} + \binom{6}{1} + \binom{6}{2} + \binom{6}{3} + \binom{6}{4} + \binom{6}{5} + \binom{6}{6}$$

SOLUTION

We know that

$$(a+b)^6 = \binom{6}{0}a^6b^0 + \binom{6}{1}a^5b^1 + \binom{6}{2}a^4b^2 + \binom{6}{3}a^3b^3 + \binom{6}{4}a^2b^4$$

$$+ \binom{6}{5}a^1b^5 + \binom{6}{6}a^0b^6$$

By substituting $a = 1$ and $b = 1$ we get

$$(1+1)^6 = \binom{6}{0}1^6 1^0 + \binom{6}{1}1^5 1^1 + \binom{6}{2}1^4 1^2 + \binom{6}{3}1^3 1^3$$

$$+ \binom{6}{4}1^2 1^4 + \binom{6}{5}1^1 1^5 + \binom{6}{6}1^0 1^6$$

Thus,

$$\binom{6}{0} + \binom{6}{1} + \binom{6}{2} + \binom{6}{3} + \binom{6}{4} + \binom{6}{5} + \binom{6}{6} = 2^6 = 64$$

### Example 37

What is the coefficient of $x^4 y^2$ in $(x+y)^6$?

SOLUTION

Using the binomial theorem we can see that the coefficient will be $\binom{6}{4} = 15$.

### Example 38

What is the coefficient of $x^3 y^2$ in $(3x - y)^5$?

SOLUTION

Using the binomial theorem we know that the coefficient of $a^3 b^2$ in the expansion of $(a+b)^5$ is $\binom{5}{3} = 10$. Now, put $a = 3x$ and $b = -y$. Then,

$$10a^3 b^2 = 10(3x)^3(-y)^2 = 10 \times 3^3 \times (-1)^2 x^3 y^2 = 270 x^3 y^2$$

So, the coefficient of $x^3 y^2$ is 270.

### EXERCISES ON 4.5

1. What is the coefficient of $x^2 y^4$ in $(x+y)^6$?
2. What is the coefficient of $x^2 y^4$ in $(2x+y)^6$?
3. What is the coefficient of $x^2 y^4$ in $(x+y-1)^7$?
4. Expand $(x+y)^7$. Then substitute $x = 1$ and $y = 1$ to obtain the value of $\binom{7}{0} + \binom{7}{1} + \binom{7}{2} + \binom{7}{3} + \binom{7}{4} + \binom{7}{5} + \binom{7}{6} + \binom{7}{7}$.
5. Expand $(x+y)^7$. Then substitute $x = 1$ and $y = -1$ to obtain the value of $\binom{7}{0} - \binom{7}{1} + \binom{7}{2} - \binom{7}{3} + \binom{7}{4} - \binom{7}{5} + \binom{7}{6} - \binom{7}{7}$.

## 4.6 Multinomials and rearrangements

How many rearrangements are there of the letters in the name ROSALIND? There are 8 letters, so we can pick the first one in 8 ways, the next in 7 ways and so on until the last letter which can be chosen in only one way. This makes 8! different rearrangements. For example, swapping the second and sixth letter produces the 'word' RISALOND.

How many rearrangements are there of the letters in NANNA? There are 5 letters in NANNA, but the letters N and A are repeated. Because we cannot distinguish between the A's (or between the N's for that matter) there are no longer 5! different rearrangements. For example, if we list all of the 5! arrangements, the arrangement NANNA will appear several times. We need an A in the 2nd and 5th places, but this can happen in 2! = 2 different ways. We need an N in the 1st, 3rd and 4th places, but this can happen in 3! different ways. Thus, the particular arrangement NANNA will appear 2!3! times. By the same argument, every different arrangement will appear 2!3! times. Thus, the number of distinguishable rearrangements is 5!/2!3! = 10.

## Example 39

How many different rearrangements are there of the letters in the word MATHEMATICS?

### SOLUTION
In MATHEMATICS the letters M, A and T are repeated, so there are no longer 11! different rearrangements because we cannot distinguish between the A's, or M's, or T's. For example, if we list all of the 11! arrangements, the arrangement MATHEMATICS will appear several times. We need an M in the 1st and 6th places, but this can happen in 2! different ways. We need an A in the 2nd and 7th places, but this can happen in 2! different ways. We need a T in the 3rd and 8th places, but this can happen in 2! different ways. Thus, the particular arrangement MATHEMATICS will appear 2!2!2! times. By the same argument, every different arrangement will appear 2!2!2! times. The letter H only appears once so we could think of this as in 1! different ways and similarly for E, I, C and S. Thus, the number of rearrangements is 11!/2!2!2!1!1!1!1!1!1! = 4 989 600. In formulae like this, we usually do not bother to write out the 1!'s.

## Example 40

How many different rearrangements are there of the word REARRANGEMENT?

### SOLUTION
There are 13 letters, so if they were all different we would have 13! ways. But the letter A is repeated twice, E is repeated three times, N is repeated twice, R is repeated three times and the other letters occur just once. Using the same argument as in the previous example, we conclude that there are 13!/2!2!3!3! = 43 243 200 different rearrangements. This is a surprisingly large number!

## Example 41

In the city whose street layout is shown in Fig 4.6, how many different ways are there of travelling from A to B if you can only move in the directions due north or due east?

### SOLUTION
Trace out the route NNNEENEEE. Work out two more possible routes. You will discover that any feasible route always has length 9, and consists of four N's and five E's. Now write down any sequence consisting of four N's and five E's. Is this a route from A to B? You will find that any such sequence defines a route. Thus, the number of ways of travelling from A to B is the same as the number of rearrangements of the 'word' NNNEENEEE, and this is 9!/4!5!.

We can express this result as follows.

**Fig 4.6**

## ● *Theorem 6*

**The multinomial theorem.** The number of rearrangements of $n_1$ objects of type 1, $n_2$ objects of type 2, ..., $n_k$ objects of type $k$, where $n = n_1 + n_2 + \ldots + n_k$, is

$$\frac{n!}{n_1! n_2! \ldots n_k!}$$

The number $n!/n_1! n_2! \ldots n_k!$ is called the **multinomial coefficient** and is written as

$$\binom{n}{n_1, n_2, \ldots, n_k}$$

## ● *Example 42*

How many different 'words' can you get from rearranging the letters in the word RECURRENCE?

SOLUTION

Some of the letters are repeated so we need to use the multinomial theorem. The word contains 3 R's, 3 E's, 2 C's, 1 N and 1 U. So there are

$$\binom{10}{3, 3, 2, 1, 1} = 10!3!3!2!1!1! = 50400$$

different words.

### Example 43

I have to stick three 1p stamps and two 10p stamps on an envelope. Assuming that I stick the stamps in a line along the top of the envelope, how many different arrangements are there?

SOLUTION
We can again use the multinomial theorem to get the solution, which is $\binom{5}{2,3} = \frac{5!}{3!2!} = 10$. Notice that $\binom{5}{2}$ is also equal to 10. The explanation is the following. Because there are only two different types of stamp we could have treated the problem as an unordered selection of 2 from 5 without repetition, since this corresponds to the number of ways of putting two 10p stamps in an envelope with 5 spaces. Then, the other three spaces must get the three 1p stamps. This gives the answer as $\binom{5}{2} = \frac{5!}{2!(5-2)!} = 10$.

Remember that we used the combination $\binom{n}{r}$ to give the coefficients of $(x+y)^n$. In this next example, we see how the multinomial $\binom{n}{r_1, r_2, \ldots, r_k}$ can be used to give the coefficients of expressions like $(x+y+z)^7$, $(a+b+c+d+e)^9$, and so on.

### Example 44

What is the coefficient of $x^2 y^3 z^2$ in $(x+y+z)^7$?

SOLUTION
This is the same as asking how many ways we can choose an $x$ from two brackets, a $y$ from three brackets and a $z$ from two brackets in the expansion

$$(x+y+z)(x+y+z)(x+y+z)(x+y+z)(x+y+z)(x+y+z)(x+y+z)$$

This is the same as asking how many rearrangements there are of the letters $xxyyyzz$. By the multinomial theorem, this is $\binom{7}{2,3,2} = \frac{7!}{2!3!2!} = 210$. Thus, the coefficient of $x^2 y^3 z^2$ in $(x+y+z)^7$ is 210.

## EXERCISES ON 4.6

1. How many ways can you rearrange the letters of ABRACADABRA?
2. How may ways can you rearrange the letters of MISSISSIPPI if the two P's must always stay together?
3. A coin is tossed 20 times. How many ways are there of getting 15 heads and 5 tails?
4. Fifty students are to be allocated accommodation in 5 different houses. House H1 has 6 rooms, houses H2 and H3 have 10 rooms, house H4 has 11 rooms and house H5 has 13 rooms. How many different ways can the students be housed?
5. Write out the expansion of $(x+y+z)^5$.
6. What is the coefficient of $x^3 y^2 z^3$ in $(x+y+z)^8$?
7. What is the coefficient of $x^2 y^3$ in $(x+y+2)^6$?
8. What is the coefficient of $x^2 y^4$ in $(x+y-1)^7$? (You have already attempted this exercise in an earlier section, using the binomial theorem—notice how the use of the multinomial makes the problem easier to solve.)
9. What is the coefficient of $x^2 y^3 z^2$ in $(x+y+z-1)^{10}$?

10. Twenty people are called for jury service. Four of them are students, seven are waged employees, five are self-employed and four are retired.
    (a) How many different twelve-person juries can be selected?
    (b) How many different *compositions* of juries can be selected? (Two juries have the same composition if they have the same numbers of students, waged employees, self-employed and retired people.)

# 4.7 Projects

## Yale keys

Look at some Yale keys and decide how many different positions and depths they use. Do you think that all of these make reasonable keys? Would you feel secure with a key that had all the positions cut to the same depth? Or no cuts at all? Make an estimate of the number of different kinds of acceptable Yale keys that can be produced.

## Poker hands

In the game of poker, a player is dealt a hand of five cards from a standard, 52-card pack. How many different possible hands are there? The 'interesting' hands are the following:

- a pair—two cards with the same face-value, e.g. a pair of threes
- two pairs—e.g. a pair of threes *and* a pair of nines
- three of a kind—three cards with the same face value, e.g. three sevens, three kings
- four of a kind—four cards with the same face value, e.g. four sevens, four kings,
- five of a kind (also known as cheating!)
- a full house—three of a kind *and* a pair
- a flush—five cards of the same suit, e.g. four, seven, eight, queen and king of spades
- a straight flush—five cards of the same suit in consecutive numerical order, e.g. eight, nine, ten, jack, queen of clubs
- a run—five cards in numerical order, but not all of the same suit, e.g. eight, nine and jack of clubs, ten of hearts, queen of diamonds.

How many different possible hands are there of each type? Notice that for the straight flush and the run, you will need to decide whether aces can be high, low or either, and whether you allow runs like queen, king, ace, two, three. Try different variations to see how much difference it makes to your answers. What would be a sensible 'ranking' of the different types of hand, i.e. which kinds of hand should beat which other kinds? Does this correspond to the ranking actually used in poker? Invent some other 'interesting' hands and work out how many of them there are. For example (with an appropriate numerical coding for jack, queen and king if you like),

- prime numbers only,
- powers of two.

# *Summary*

**Rule of Sum.** If a set of objects can be divided into $m$ disjoint subsets and the $i$th subset has $n_i$ elements, then we have exactly $n_1 + n_2 + \ldots + n_m$ objects altogether.

**Rule of Product.** If an activity can be composed from $m$ different objects, of which the $i$th can be chosen in $n_i$ ways, then we have exactly $n_1 \times n_2 \times \ldots \times n_m$ ways of composing the activity.

**Selections of $n$ objects from $k$:**

|  | Repetition | No repetition |
|---|---|---|
| Ordered | $n^k$ | $\frac{n!}{(n-k)!} = (n)_k$ |
| Unordered | $\binom{n-1+k}{k}$ | $\frac{n!}{(n-k)!k!} = \binom{n}{k}$ |

An ordered selection without repetition is a **permutation**. An unordered selection without repetition is a **combination**.

**Pascal's triangle:**

$$
\begin{array}{ccccccccccc}
 & & & & & 1 & & & & & \\
 & & & & 1 & & 1 & & & & \\
 & & & 1 & & 2 & & 1 & & & \\
 & & 1 & & 3 & & 3 & & 1 & & \\
 & 1 & & 4 & & 6 & & 4 & & 1 & \\
1 & & 5 & & 10 & & 10 & & 5 & & 1 \\
\end{array}
$$

The entries in the triangle correspond to the numbers $\binom{n}{k}$. The $k$th position in the $n$th row contains $\binom{n}{k}$.

**The binomial theorem.** If $a$ and $b$ are real numbers and $n$ is a positive integer, then

$$(a+b)^n = \binom{n}{0}a^n b^0 + \binom{n}{1}a^{n-1}b^1 + \binom{n}{2}a^{n-2}b^2 + \ldots + \binom{n}{n}a^0 b^n$$

The number of **rearrangements** of $n_1$ objects of type 1, $n_2$ objects of type 2, $\ldots$, $n_k$ objects of type $k$, where $n = n_1 + n_2 + \ldots + n_k$, is

$$\frac{n!}{n_1! n_2! \ldots n_k!}$$

The number $\frac{n!}{n_1! n_2! \ldots n_k!}$ is called the **multinomial coefficient** and is written as

$$\binom{n}{n_1, n_2, \ldots, n_k}$$

# 5 • Probability

Letter to *The Guardian*, January 1993.

> For a £650 non-refundable fee to the London Gender Clinic a couple can have more than a 60 per cent chance of giving birth to a child of the desired sex. But couples should not overlook the fact that under normal circumstances the chance of giving birth to a child of the desired sex is 50 per cent with no fee payable.
>
> An alternative method is available for increasing this probability. A couple with a gender preference should plan to have two children. That way, the chance of having at least one child of the desired sex is 75 per cent. A couple planning three children have an 87.5 per cent chance of this, and a couple planning four have a 93.75 per cent chance: and all perfectly natural.
>
> (Dr) Peter G. Moffatt

How did Dr Moffatt derive his figures?

In this chapter we shall show how the ideas we have developed so far in the book can be used in *probability*, which is the branch of mathematics dealing with phenomena whose outcomes involve uncertainty—like the sex of an as-yet-unborn child. We first consider the special case in which the possible outcomes are all equally likely, before going on to consider general sets of outcomes. We then develop rules for combining probabilities which turn out to have very strong connections with the rules for manipulating sets which we discussed in Chapter 2. Finally, we describe the important ideas of conditional probability and independence, and show you how to use these ideas to solve some quite complicated problems involving sequences of random outcomes.

## 5.1 Introduction

The fundamental idea in probability theory is that probability can be measured on a scale which runs from zero (representing impossibility) to one (representing certainty). In everyday life, we often make unconscious judgements about probability. For example, when you leave your house in the morning, you may decide to take a raincoat with you, even if it is not raining, because you think it might rain later in the day—you judge that there is a non-zero *probability* of rain. If you are interested in politics, you might be prepared to say things like 'I think that Labour has a better than even chance of winning the next General Election'. How can we develop a mathematical theory to explain this kind of reasoning?

The origins of probability theory are often associated with two French mathematicians, Pascal (whom we first encountered in Chapter 4) and Pierre de Fermat (1601–1665).

**Fig 5.1** Pierre de Fermat (David Eugene Smith Collection, Rare Book and Manuscript Library, Columbia University)

In 1654 Pascal started a correspondence with Fermat, who wrote most of his work not as published papers but in letters to other mathematicians.

The letters between Pascal and Fermat addressed the following problem. Some gamblers in France wanted to know how to determine the division of stakes in an interrupted game of chance between two equally skilled players, knowing the scores of the players at the time of interruption and the number of points needed to win the game. Fermat and Pascal corresponded about the problem and in their solution made use of Pascal's triangle.

Suppose that $A$ and $B$ are the two players and that when the game is interrupted, $A$ needs 2 points to win whereas $B$ needs 3 points to win. Intuitively, you would guess that had the game not been interrupted, $A$ would have been more likely than $B$ to be the eventual winner, and a 'fair' division of the stakes should recognise this. We can see that after four more games, one of the two players must have won. The 16 possible outcomes of the four extra games are

aaaa aaab aaba abaa baaa aabb abba abab bbaa baab baba

all of which would result in a win for $A$, and

bbbb bbba bbab babb abbb

which result in a win for $B$.

So $A$ wins in 11 cases and $B$ wins in 5. Therefore the stakes from the interrupted game should be divided in the ratio 11:5. An interesting series of letters on this problem between Fermat and Pascal can be found in *Source Book in Mathematics,* by D. E. Smith, published by McGraw-Hill.

## 5.2 Equally likely outcomes

### Example I

If we flip a coin we know that either a head or a tail will occur and, unless there is something unusual about the coin, we believe that the chance of a head or a tail is the same. To think about what this statement means, suppose we flip a coin 10 times and count the number of heads and the number of tails, then repeat the experiment with 100 flips, and then with 500. The following table of results of such an experiment suggests that the proportion of heads is getting closer to $\frac{1}{2}$ the more coin flips we do.

| Number of flips | Number of heads | Number of tails | Proportion of heads |
|---|---|---|---|
| 10 | 6 | 4 | 0.6 |
| 100 | 46 | 54 | 0.46 |
| 500 | 247 | 253 | 0.494 |

There is another way of looking at the last of these three experiments. Instead of just calculating the proportion of heads in the whole 500 flips, we could look at how the proportion of heads in the first $n$ flips changes as $n$ increases. Call this proportion $p_n$. Fig 5.2 plots $p_n$ against $n$. At first, the graph fluctuates wildly. For example, because the first four flips were tail, tail, head, tail, the graph starts at $p_1 = 0$ and $p_2 = 0$, then increases to $p_3 = \frac{1}{3}$, decreases again to $p_4 = \frac{1}{4}$, and so on. But after the initially erratic behaviour, it settles down to a proportion close, but not exactly equal, to $\frac{1}{2}$.

Fig 5.3 shows a repetition of the same experiment. This time, the first four flips gave tail, head, head, head, so the graph starts off with $p_1 = 0$, $p_2 = \frac{1}{2}$, $p_3 = \frac{2}{3}$, $p_4 = \frac{3}{4}$ and so on. The initial part of the graph is again erratic, and looks quite

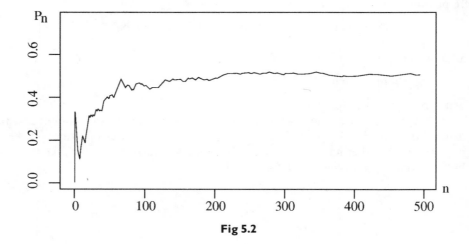

**Fig 5.2**

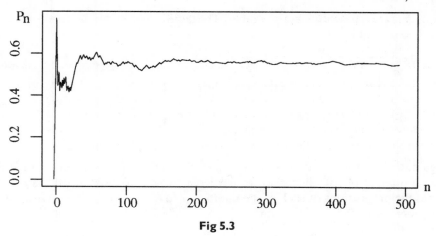

**Fig 5.3**

different from the initial part of the previous graph. But eventually, it too settles down to a proportion close to $\frac{1}{2}$.

Loosely speaking, the proportion of heads that we would get if we did the experiment a very large number of times is called the *probability* of a head. Our experiment suggests that the probability of a head is $\frac{1}{2}$. The probability of a tail must also be $\frac{1}{2}$, since the proportions of heads and tails must add to 1. Of course this assumes that our coin is perfectly balanced, i.e. a *fair* coin. The coin might have been unbalanced so that it landed fewer times on heads than on tails. Figure 5.4 shows the result of an experiment with a coin which has probability 0.4 of landing heads.

We could try a similar experiment with a die.

## Example 2

We roll the die and note whether we get a 1, 2, 3, 4, 5 or 6. Assuming that our die is perfectly symmetrical, we would have no reason to suspect that any number would

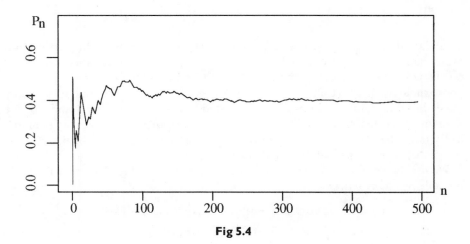

**Fig 5.4**

turn up more frequently than any other. The table below shows the results for 6, 60 and 600 rolls of the die.

| $n$ | 1 | 2 | 3 | 4 | 5 | 6 |
|---|---|---|---|---|---|---|
| 6 | 0 | 1 | 3 | 0 | 1 | 1 |
| 60 | 8 | 10 | 18 | 12 | 6 | 6 |
| 600 | 99 | 97 | 102 | 103 | 105 | 94 |

Again, we see rather erratic results from a small number of rolls, but a clear pattern as the number of rolls increases. In particular, when we have rolled the die 600 times the proportions of each of the numbers 1 to 6 are all close to $\frac{1}{6}$. We say that the *probability* of each of the numbers 1, 2, 3, 4, 5 or 6 is $\frac{1}{6}$.

## Example 3

I have three keys in my pocket and I pull one out to unlock my office door. Assuming that each key is equally likely to be pulled out first and only one fits the lock on my office, how likely is it that I will get the right key first time?

SOLUTION

Since there are three keys, the first key will be the right one one-third of the times. Hence the probability of opening my office with the first key that I pick is $\frac{1}{3}$.

## Example 4

Suppose you have a fair tetrahedron (so it is equally likely to land on any one of its four faces) with sides coloured red, blue, green, yellow. If you throw it 100 times how many times would you expect it to land on its red face? What is the probability of a red on each throw?

SOLUTION

You would expect 25 red throws out of 100. The probability of a red is therefore $25/100 = \frac{1}{4}$.

You should realise that in this last example we are using the word 'expect' in a rather subtle way. We would actually be quite surprised if the tetrahedron landed on its red face *exactly* 25 times—in fact, the probability of this happening is approximately 0.092. (Can you prove this? If not, you should be able to do so by the end of this chapter!) But if you were forced to make a guess in advance for the number of red faces, 25 would be the most sensible guess—in fact, it is the most likely outcome of the experiment in the sense that any number other than 25 has associated with it a probability smaller than 0.092.

We use the word **experiment** to describe any procedure which has more than one possible result, or **outcome**.

The set of outcomes for the coin-flipping experiment is {head, tail}.

The set of outcomes for the die-rolling experiment is {1, 2, 3, 4, 5, 6}.

Each outcome has an associated probability, which corresponds to the proportion of times we would expect to get that particular outcome if we repeated the experiment a large number of times. We write the probability of the outcome as $P(\cdot)$, and in the brackets put something to show which outcome is being described. In Example 1, the coin-flipping experiment, we would use H and T to represent heads and tails, to give

$$P(H) = \tfrac{1}{2}, P(T) = \tfrac{1}{2}$$

In Example 2, the die-rolling experiment,

$$P(1) = P(2) = P(3) = P(4) = P(5) = P(6) = \tfrac{1}{6}$$

## EXERCISES ON 5.2

1. The astragalus described in Example 31 of Chapter 4 will rest on one of four sides with equal probabilities. What is the probability of the bone landing on any particular side?

2. In the following experiments, decide what are the outcomes and state if they are equally likely:
   (a) A red, a blue and a yellow ball are put into a bag and a ball drawn out at random.
   (b) A record is taken of the sex of all births in a year in a maternity ward.
   (c) Two children throw a die to see who starts the game. If the die shows a prime number Joe starts, otherwise Ellen starts.

3. Match the following probabilities (i) to (v) with the statements (a) to (e). (i) 0; (ii) 0.03; (iii) 0.6; (iv) 0.99; (v) 1.
   (a) The event is impossible and can never occur.
   (b) The event is certain and will occur on every trial.
   (c) The event is very unlikely but will occur once in a while in a long run of trials.
   (d) The event will occur more often than not.
   (e) The event will almost always occur.

4. You are given a fair 10-sided spinner numbered $0, 1, \ldots, 9$. What is the probability of throwing a 5? How could you use the spinner to get all values between 0 and 99 with equal probabilities $1/100$?

5. Suppose you are gambling with a fair coin and you can choose to do either 10 or 100 flips.
   (a) In the first game you will win if the proportion of heads is between 0.4 and 0.6. Are you more likely to win if you choose 10 or 100 flips?
   (b) In the second game you will win if exactly half of the flips are heads. Are you more likely to win if you choose 10 or 100 flips?

6. From the design of a coin it looks obvious that in a large number of flips it will fall heads roughly half the time and experiment shows this to be true. If you hold a coin on its end on a hard surface and flick it from the side so it spins, are the probability of a head and that of a tail still equal? Do some experiments of your own to find out.

# 5.3 Experiments with outcomes which are not equally likely

In the examples of the previous section, the probability of each possible outcome was the same. This is not always the case, as the following example shows.

## ● *Example 5*

A pair of dice are thrown. If we add the numbers on the two dice, what are the possible sums? The sums can range from 2 to 12, but are these sums all equally likely to occur? What is the probability that the sum of the two numbers on the dice is 7?

SOLUTION

If the sums were equally likely then we would expect each sum to have probability 1/11 since there are 11 possible sums. Here are the results of 1100 repetitions of the experiment.

| Sum | Number of occurrences |
|-----|-----------------------|
| 2   | 34                    |
| 3   | 67                    |
| 4   | 93                    |
| 5   | 126                   |
| 6   | 158                   |
| 7   | 182                   |
| 8   | 156                   |
| 9   | 110                   |
| 10  | 93                    |
| 11  | 61                    |
| 12  | 20                    |

The table makes it clear that the observed proportions are rather different from each other, so it seems unlikely that the different outcomes are all equally likely.

If we look more closely at what happens when we throw two dice, we can argue as follows. The first die is equally likely to give any number between 1 and 6. So is the second die. But the different possible values for the sum of the two numbers are not equally likely because, for example, the only way you can get a sum of 2 is to throw a 1 on both dice, whereas a 7 can arise from a 1 and 6, a 2 and 5 or a 3 and 4. The complete story is told in the following table, in which the numbers in the body of the table are the sums obtained from the different possible results for each die.

|  |  | Die 1 | | | | | |
|---|---|---|---|---|---|---|---|
|  |  | 1 | 2 | 3 | 4 | 5 | 6 |
| Die 2 | 1 | 2 | 3 | 4 | 5 | 6 | 7 |
|  | 2 | 3 | 4 | 5 | 6 | 7 | 8 |
|  | 3 | 4 | 5 | 6 | 7 | 8 | 9 |
|  | 4 | 5 | 6 | 7 | 8 | 9 | 10 |
|  | 5 | 6 | 7 | 8 | 9 | 10 | 11 |
|  | 6 | 7 | 8 | 9 | 10 | 11 | 12 |

We can see that the sum 7 occurs six times whereas the sum 2 only occurs once. So we would not expect our proportions to be the same. In fact, we would expect the proportion of sum 7 to be about $6/36 = 1/6$, which is the probability, i.e. $P(7) = 1/6$. Similarly, we expect the proportion of sum 2 to be about $1/36$, and in fact $P(2) = 1/36$. We can work out all of the probabilities from the numbers of occurrences of each sum in our table and see how these compare with the values we obtained by rolling a pair of dice 1100 times.

| Sum | Number of occurrences | Observed proportion | Theoretical probability |
|---|---|---|---|
| 2 | 34 | 0.031 | 0.028 |
| 3 | 67 | 0.061 | 0.056 |
| 4 | 93 | 0.085 | 0.083 |
| 5 | 126 | 0.115 | 0.111 |
| 6 | 158 | 0.144 | 0.139 |
| 7 | 182 | 0.165 | 0.167 |
| 8 | 156 | 0.142 | 0.139 |
| 9 | 110 | 0.100 | 0.111 |
| 10 | 93 | 0.085 | 0.083 |
| 11 | 61 | 0.055 | 0.056 |
| 12 | 20 | 0.018 | 0.028 |

Another way to find the probabilities for the sum of two dice without having to tabulate all the possibilities uses the sum and product rule from §4.2.

From our dice table we can see that there are 36 possible outcomes for the two dice, only one of which is the outcome $(1, 6)$, i.e. 1 on the first die and 6 on the second. So the probability of $(1, 6)$ is $1/36$. Another way to derive this is to use the rule of product. We know that a 1 occurs on the first die with probability $1/6$ and a 6 occurs on the second die with probability $1/6$. Since we want both of these outcomes to happen we use the multiplication rule to see that the outcome $(1, 6)$ happens with probability $1/6 \times 1/6 = 1/36$.

However there are other ways to get a sum of 7, i.e. $(2, 5), (3, 4), (4, 3), (5, 2)$ and $(6, 1)$. Each of these outcomes has probability $1/36$. Since any one of these gives us a sum of 7 we now use the rule of sum to add the individual probabilities, giving

$$1/36 + 1/36 + 1/36 + 1/36 + 1/36 + 1/36 = 6/36 = 1/6$$

You could use a similar argument to obtain the probability of a sum of 11. The possible dice throws are (5,6) and (6,5) and these each have probability $1/6 \times 1/6 = 1/36$, hence the probability of a sum of 11 is $1/36 + 1/36 = 2/36 = 1/18$. We can use the same ideas in other problems.

## Example 6

In the card game pontoon, each player is dealt two cards. To claim *pontoon* the player's two cards must add to 21. The picture cards all have value 10, the aces have value 11 and the rest of the cards have face value. What is the probability of getting *pontoon*?

SOLUTION
From results in §4.3, we see that there are $\binom{52}{2}$ different two-card hands, since the two cards are an unordered choice of 2 from 52 without repetition. To see how many of these add up to 21 we need to list all possible pairs of cards adding to 21. With two cards, the only way to get 21 is with an 11 and a 10. There are 4 aces and 16 cards of value 10, namely the 4 kings, 4 queens, 4 jacks and 4 tens. By the multiplication rule there are $4 \times 16$ possible pairs. Hence, the probability of pontoon is

$$\frac{64}{\binom{52}{2}} = 0.048\ 265$$

Now, you might argue that in each of these examples, the experiment *does* have equally likely outcomes—we have just grouped different numbers of them together. So, in the first example, you could say that there are 36 outcomes, of which only one gave a sum of 2, whereas six of the 36 outcomes gave a sum of 7. Similarly, in the second example all possible pairs of cards are equally likely, and 64 different pairs all give *pontoon*. Another way of putting this is that the *outcomes* of the experiment are equally likely, but we are only interested in specified *sets of outcomes*, called **events**. So, in the first example, the events of interest correspond to the 11 different possible values of the sum, and in the second example the event of interest is *pontoon*.

In other situations, the outcomes of the experiment have fundamentally different probabilities. Here is a famous example from the early history of probability.

## Example 7

In 1777 Comte de Buffon (1707–1788), director of the Paris Jardin du Roi, devised the following experiment to determine $\pi$ by a probabilistic method. The experiment involves randomly dropping a needle onto a surface marked with parallel lines all spaced the same distance apart as the needle's length, and counting the proportion of times the needle falls across a line. Buffon showed that the probability of the needle falling across a line would be $2/\pi$. Fig 5.5 is an old drawing of this experiment.

Fig 5.6 shows a simulation of this experiment with 500 throws of the needle. As in Example 1, the initial part of the graph is erratic, but eventually it settles down to a proportion close to $2/\pi$.

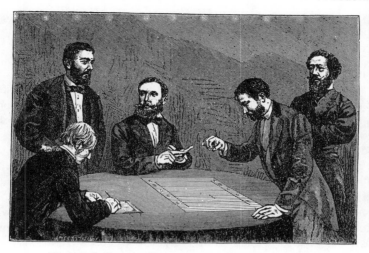

**Fig 5.5** Buffon needle experiment (source unknown)

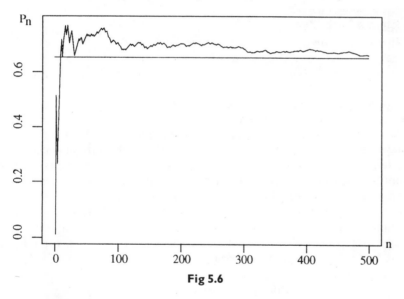

**Fig 5.6**

Here is a challenge. Can you *prove* that the probability that the needle crosses the line is $2/\pi$? You will need to assume that the centre of the needle is equally likely to land anywhere between two adjacent parallel lines, and that the orientation of the needle is equally likely to be in any direction. Let the needle be of length 1, let $x$ denote the perpendicular distance from the nearest line to the centre of the needle (so $x$ lies between 0 and $\frac{1}{2}$), and let $\theta$ denote the orientation of the needle relative to the parallel lines (so $\theta$ lies between $-\pi/2$ and $\pi/2$ radians). Draw diagrams which show the needle not crossing, just touching, and crossing the line. Determine the ranges of values for $x$ and $\theta$ for which the needle crosses the line (noting that the allowable range of $\theta$ depends on $x$ and vice versa). Think of each possible pair $(x, \theta)$ as defining a point in a rectangle of height $\frac{1}{2}$ (because $0 \le x \le \frac{1}{2}$)

and width $\pi$ (because $-\pi/2 \le \theta \le \pi/2$). What proportion of the area of this rectangle corresponds to points $(x, \theta)$ for which the needle crosses the line?

To take another example, if a die is not perfectly symmetrical, then the proportions of times we get the numbers 1 to 6 in repeated rolls of the die will not be the same. But we can still talk about the probability of a 1, a 2, and so on. What properties must these probabilities have? Since we want them to correspond to long-run proportions, they must be non-negative and their sum must be 1.

This idea is fairly intuitive, but it was only in the twentieth century that the Russian probabilist, Andrei Nikolaevich Kolmogorov (1903–1987), provided a rigorous mathematical framework to discuss probabilities for experiments whose outcomes are not all equally likely.

In the next section, we develop a more formal language of probability which will allow us to devise rules for working out probabilities in more complicated situations.

### EXERCISES ON 5.3

1. What is the probability that when three coins are flipped there will be exactly three heads? One tail and two heads?
2. What is the probability of drawing an ace from each of the following:
   (a) a standard 52 card pack?
   (b) a pack consisting only of red cards?
   (c) a pack consisting only of picture cards?
3. Many American states run lotteries. In the original New York lottery the first prize was $50 000, there were 9 prizes of $5000, 90 prizes of $500 and 900 prizes of $50. Altogether, 1 000 000 tickets were sold at 50 cents each. What percentage of the ticket income was paid out in prizes? How does this lottery compare with casinos which pay out about 90% of their income?
4. A netball player scores from 75% of her goal shots in all the practice games. In the first tournament game she shoots for the goal five times and misses three times. Her coach tells her this is because she is nervous ... but could the misses have just been due to chance? What is her probability of missing at least three out of five shots?
5. If you assume that the probability of giving birth to a boy is the same as for a girl, then if a person already has five girls what is the probability that they will have a girl next time? What is the probability of a person with six children having six girls? Why are the two answers different?

## 5.4 The sample space, outcomes and events

In §5.2 and §5.3 we have seen many examples of sets of outcomes. We call the set of all possible outcomes of an experiment the **sample space**, $S$.

For the coin-flipping example, $S = \{H, T\}$.

For the tetrahedral die, $S = \{\text{red, blue, green, yellow}\}$.

The elements of a sample space are called **elementary outcomes**, often abbreviated to **outcomes**. These must all be disjoint—when the experiment is performed, one and only one of the outcomes occurs. We write the outcomes as $e_1, e_2, \ldots$. In this book, we will generally consider only experiments for which the number of elementary outcomes is either finite or countably infinite. (Buffon's needle is an exception—in that experiment, the elementary outcomes consist of all possible locations and orientations of the needle, corresponding to a non-countably infinite sample space.)

For example, if the experiment consists of flipping a single coin, the sample space is

$$S = \{e_1, e_2\}$$

where $e_1$ =head and $e_2$ =tail. So $S$ has a finite number of elements.

On the other hand, if the experiment consists of counting the number of times you need to flip a coin before you get a tail, the sample space is

$$S = \{e_1, e_2, \ldots\}$$

where $e_i$ corresponds to a sequence of $i - 1$ heads followed by a tail. In this case, $i$ can be any positive integer (even if very large values of $i$ are rather unlikely!), and $S$ has an infinite number of elements.

An **event** is any subset of $S$. Pairs of events may or may not be disjoint. For example, in the tetrahedral die of Example 4, two events might be {red, blue} and {red, green}, which are not disjoint. Notice that the outcomes, $e_i$, are also events because they are (single-element) subsets of $S$.

Let $E$ be any event, and $e$ the elementary outcome which we observe. If $e \in E$, we say that the event $E$ *has occurred*. If $e \notin E$, we say that $E$ *has not occurred*.

We use the notation and terminology of set theory to build complicated events from simple ones. Thus,

- $A^c$ is the **complement** of $A$. This means that $A$ does not occur.
- $A \cup B$ is the **union** of $A$ and $B$. This means that either $A$ occurs, or $B$ occurs, or both.
- $A \cap B$ is the **intersection** of $A$ and $B$. This means that both $A$ and $B$ occur.

We have seen that each outcome has a probability associated with it. So, for example, for the coin flipping example,

$$P(H) = \tfrac{1}{2}, P(T) = \tfrac{1}{2}$$

and in the tetrahedral die example,

$$P(\text{red}) = \tfrac{1}{4}, P(\text{green}) = \tfrac{1}{4}, P(\text{blue}) = \tfrac{1}{4}, P(\text{yellow}) = \tfrac{1}{4}$$

As we predicted at the end of §5.3, in each example the probabilities are all non-negative numbers, and the sum of the probabilities corresponding to all of the elements of the sample space is 1.

To make this rigorous, we use 'capital letters, $A$, $B$, etc. to denote *events*, and write $P(A)$ for the *probability* of the event $A$. Then, the **probability function** $P(.)$ is a mapping from the sample space $S$ to the real numbers which obeys the following axioms:

## Kolmogorov's axioms

*Axiom 1*: For any event $A$, $P(A) \geq 0$.
*Axiom 2*: $P(S) = 1$.
*Axiom 3*: If $A$ and $B$ are disjoint events, then $P(A \cup B) = P(A) + P(B)$.

Throwing two dice in Example 5 gives us an illustration of Axiom 3. In that example, we considered the event $E$ of throwing two dice whose sum was 7. The event $E$ is composed of the outcomes $e_1 = (1, 6)$, $e_2 = (2, 5)$, $e_3 = (3, 4)$, $e_4 = (4, 3)$, $e_5 = (5, 2)$ and $e_6 = (6, 1)$, so

$$P(E) = P(e_1 \cup e_2 \cup \ldots \cup e_6)$$

Now, because the $e_i$ are outcomes they are disjoint and Axiom 3 states that

$$P(E) = P(e_1) + P(e_2) + \ldots + P(e_6)$$

which we can write as

$$P(E) = \sum_{i=1}^{6} P(e_i) = \sum_{e \in E} P(e)$$

Strictly we are being a bit sloppy with our notation here, because we are not distinguishing between the *element* $e_i$ and the single-element *set* $\{e_i\}$.

We can now use the axioms to prove some fundamental results about probabilities. Firstly, the example above can be generalised as the following theorem.

## ● *Theorem 1* ─────────────────────

The probability of any event $A$ is the sum of the probabilities of all outcomes $e_i \in A$.

PROOF
Let $A = \{e_1, e_2, \ldots, e_n\}$ and, for $i = 2, \ldots, (n-1)$ define $A_i = \{e_1, \ldots, e_i\}$. Then, $A_2 = \{e_1\} \cup \{e_2\}$, and since $\{e_1\}$ and $\{e_2\}$ are disjoint we can use Axiom 3 directly to give $P(A_2) = P(e_1) + P(e_2)$. Now, write $A_3 = A_2 \cup \{e_3\}$, note that $A_2$ and $\{e_3\}$ are disjoint, and use Axiom 3 again, to give

$$P(A_3) = P(A_2) + P(e_3) = P(e_1) + P(e_2) + P(e_3)$$

Continuing in this way, we eventually get

$$P(A_{n-1}) = P(e_1) + \ldots + P(e_{n-1})$$

and

$$P(A) = P(A_{n-1}) + P(e_n) = P(e_1) + \ldots + P(e_n)$$

as required.

Secondly, we noticed in §5.2 that the sum of the probability of an event and the probability of its complement was 1. We shall now use our axioms to prove this more formally.

## • *Theorem 2*

Let $E$ be any event in our sample space $S$. Then

$$P(E^c) = 1 - P(E)$$

PROOF

Suppose that $E$ is the set of outcomes $\{e_1, e_2, \ldots, e_i\}$ and let $S = \{e_1, e_2, \ldots, e_n\}$. Then $E^c$ is the set of outcomes $\{e_{i+1}, e_{i+2}, \ldots, e_n\}$. So $E$ and $E^c$ have no outcomes in common. Since $S = E \cup E^c$ and $P(S) = 1$ by Axiom 2, then $P(E \cup E^c) = 1$. Hence

$$1 = P(E \cup E^c) = P(E) + P(E^c)$$

by Axiom 3, and

$$P(E^c) = 1 - P(E)$$

as required.

As a corollary to this result, we can present the following theorem.

## • *Theorem 3*

For any event $A$,

$$0 \leq P(A) \leq 1$$

PROOF

The left-hand inequality is just Axiom 1. For the right-hand inequality, write

$$P(A) = 1 - P(A^c) \leq 1$$

since $P(A^c) \geq 0$ by Axiom 1.

Another important result is Theorem 4.

## • *Theorem 4*

$$P(A \cup B) = P(A) + P(B) - P(A \cap B)$$

PROOF

We can prove this result by considering the outcomes which go to make up the events involved. A Venn diagram (Fig 5.7) helps to make the argument clear. In the diagram, the individual outcomes in the sample space are marked as solid dots. As in our work on set theory, we see that

$$|A \cup B| = |A| + |B| - |A \cap B|$$

because if we simply add the numbers of elements in $A$ and in $B$, those in $A \cap B$ get

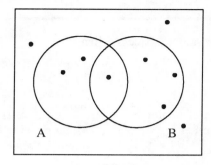

**Fig 5.7**

counted twice. Hence,

$$P(A) + P(B) = \sum_{e_i \in A} P(e_i) + \sum_{e_i \in B} P(e_i)$$

$$= \sum_{e_i \in A \cup B} P(e_i) - \sum_{e_i \in A \cap B} P(e_i)$$

$$= P(A \cup B) - P(A \cap B)$$

which gives us our result

$$P(A \cup B) = P(A) + P(B) - P(A \cap B)$$

For three sets we have the result that

$$P(A \cup B \cup C) = P(A) + P(B) + P(C) - P(A \cap B) - P(A \cap C)$$
$$- P(B \cap C) + P(A \cap B \cap C)$$

and again this can be proved by using the corresponding set-theoretic argument, which is equivalent to the inclusion–exclusion principle discussed in Section 2.5. The Venn diagram in Fig 5.8 should help you to see how the various intersections need to be added and subtracted to get the right answer.

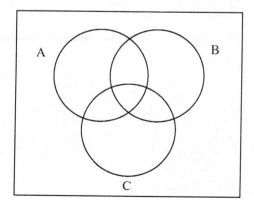

**Fig 5.8**

1. Let $S = \{a, b, c, d\}$. It is known that $P(a) = \frac{1}{4}$, $P(b) = \frac{1}{6}$, $P(c) = \frac{1}{5}$. What must be the probability of $d$, to make P a valid probability function on $S$?

2. The *New York Times* on 29 August 1989 asked, 'Does your husband do a fair share of housework?'. The replies were given as follows: more than his fair share 0.12; his fair share 0.61; and less than his fair share 0.27. Check that Kolmogorov's axioms apply to these probabilities. What is the probability that a husband does at least a fair share?

3. Given the following probabilities obtained from a large sample of men, what should the probability of being married be? The probability of being single is 0.28, widowed 0.04 and divorced 0.3.

4. Prove that

$$P(A \cup B \cup C) = P(A) + P(B) + P(C) - P(A \cap B) - P(A \cap C)$$
$$- P(B \cap C) + P(A \cap B \cap C)$$

5. A production manager declares that an acceptable quality level for plant output is that at most 1% of items produced should be faulty. To achieve this, he orders that a random sample of 100 items be taken from each day's production, and the day's production scrapped if more than one of the 100 sampled items are found to be faulty.

   Find the probabilities that a day's production is scrapped when the overall proportion of faulty items is
   (a) at the (just) acceptable level of 1%,
   (b) at the unacceptable level of 2%.
   Derive a function which gives the probability that a day's production is scrapped in terms of the overall proportion, $p$ say, of faulty items, and sketch a graph of this function.

# 5.5 Conditional probability, independence and Bayes' theorem

## Conditional probability

Medical evidence suggests that about 2.625% of the population is colour blind. This suggests that if you pick a person at random from your class, the probability that they will be colour blind is 0.026 25. However, it is also known that 5% of men are colour blind and only 0.25% of women. If you pick a person at random from your class, you can immediately tell whether they are female or male. How does this information affect the probability that they will subsequently prove to be colour blind? The probability changes to 0.0025 if you pick a woman, and to 0.05 if you pick a man. The probabilities change when some relevant partial information, in this case the person's sex, is specified. The probability of the event, $C$ say, that the person picked is colour blind, given the event, $M$ say, that a man is picked, can be written as

$$P(C|M) = 0.05$$

and similarly the probability of $C$ given the event, $W$ say, that a woman is picked is

$P(C|W) = 0.0025$

## Example 8

A friend of ours has two children. What is the probability that he has two girls? How does this probability change if I subsequently tell you that one of his children is called Jessica?

SOLUTION
If we let $G$ be the event *girl* and $B$ the event *boy*, then the sample space is

$S = \{GG, GB, BG, BB\}$

We shall assume that each of the four outcomes is equally likely. So,

$P(GG) = \frac{1}{4}$

Once we are told that our friend has a girl called Jessica, the problem changes. We now want to know the probability that he has two girls, given that he has at least one girl. We write this as

$P(GG|\text{at least one girl})$

Now, the sample space is

$S = \{GG, GB, BG\}$

because we know that our friend cannot have two boys. So the probability of two girls is $\frac{1}{3}$.

In the above example the probability changes because we have more information. However, if I tell you not that there is a child called Jessica but that there is one who goes to school, then this information does not change our initial assumption that the probability of two girls is $\frac{1}{4}$, because the information that one of the children goes to school is irrelevant to the question of whether both children are girls.

## Example 9

A coin is flipped twice. What is the probability of two heads given that the first flip was a head?

SOLUTION
This time the sample space $S$ is $\{HH, HT\}$ so

$P(HH|\text{ first flip a head }) = \frac{1}{2}$

Notice that in both of these examples, the partial information has the effect of eliminating part of the sample space, $S$. Then, to work out the new probability of the event of interest, we have to scale the probabilities of all of the outcomes which have *not* been eliminated, so that the sum of the probabilities is still 1.

The Venn diagram of Fig 5.9 illustrates this. If we know that the event $B$ has occurred, then we have reduced the original sample space $S$ to the event $B$. So, if we

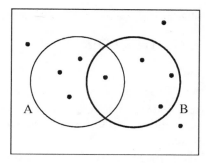

**Fig 5.9**

want the probabilities still to sum to 1, we must divide each of them by P($B$). Within $B$, the event $A$ is equivalent to the intersection, $A \cap B$. So the probability of the event $A$, given the event $B$, must be proportional to P($A \cap B$).

In the diagram, the circle drawn with the thicker black line represents the event $B$. Conditioning on $B$ is equivalent to considering only those outcomes which are inside this circle. So if, for example, all of the outcomes shown in the diagram were equally likely, conditioning on $B$ is equivalent to considering only the four outcomes in $B$ as the possible outcomes of the experiment. Of these, only one is in $A$, so the conditional probability of $A$ given $B$ is $\frac{1}{4}$.

In general, if $A$ and $B$ are any two events then

$$P(A|B) = \frac{P(A \cap B)}{P(B)}$$

Notice that this only makes sense if P($B$) > 0. If P($B$) = 0 then the event $B$ never happens and it is meaningless to ask what is the probability of $A$ happening given that $B$ happens. Continuing with Fig 5.9, we can see that if all of the outcomes are equally likely, then P($B$) = 4/10 and P($A \cap B$) = 1/10, so using the rule above, we would calculate

$$P(A|B) = \frac{P(A \cap B)}{P(B)} = \frac{1/10}{4/10} = \frac{1}{4}$$

which agrees with our first answer.

## Example 10

What is the probability that a 6 is obtained on one of the dice in a throw of two dice, given that the sum is 7?

SOLUTION

The portion of the sample space corresponding to sum 7 is the set of six outcomes

$$\{(1,6)(2,5)(3,4)(4,3)(5,2)(6,1)\}$$

Two of these have a 6 so the probability is 2/6 = 1/3.

Now let us check that we get the same answer using the general rule. The probability of one dice showing a 6 and the sum being 7 is 2/36. The probability of summing to 7 is 6/36. Hence,

$$\frac{P(A \cap B)}{P(B)} = \frac{2/36}{6/36} = 2/6 = 1/3 = P(A|B)$$

## Example 11

An experiment is being run on 100 people to test drugs $A$ and $B$ and a placebo $C$. Of the 100 patients, 30 receive drug $A$, 30 receive drug $B$ and the remaining 40 receive the placebo $C$. Given that the patient did not get drug $B$ what is the probability that they got the placebo?

SOLUTION
The general rule gives

$$P(C|B^c) = \frac{P(C \cap B^c)}{P(B^c)}$$

Now,

$$P(B^c) = 1 - P(B) = 1 - 30/100 = 7/10$$

and

$$P(C \cap B^c) = P(C) = 4/10$$

since the patient receives $C$ and does not receive $B$ if and only if the patient receives $C$. Thus

$$P(C|B^c) = \frac{4/10}{7/10} = 4/7$$

## Example 12

What is the probability of picking a club, event $C$, given that I have picked a black card, event $B$?

SOLUTION
$P(B) = 1/2$ and $P(B \cap C) = P(C) = 1/4$. So

$$P(C|B) = \frac{1/4}{1/2} = 1/2$$

Notice that in the last example the probability of picking a black card given that a club was picked is

$$P(B|C) = \frac{P(B \cap C)}{P(C)} = \frac{1/4}{1/4} = 1$$

as you would expect since if you are given the information that a club was picked, then the card is certain to be black. So we can see that in general, $P(B|C) \neq P(C|B)$. Since we have

$$P(A|B) = \frac{P(A \cap B)}{P(B)}$$

and

$$P(B|A) = \frac{P(A \cap B)}{P(A)}$$

we can rearrange the first of these to give

$$P(A \cap B) = P(A|B)P(B)$$

and rearrange the second to give

$$P(A \cap B) = P(B|A)P(A)$$

This establishes the following result.

## ● *Theorem 5*

$$P(A \cap B) = P(A|B)P(B) = P(B|A)P(A)$$

An immediate consequence of Theorem 5 is that

$$P(B|A) = \frac{P(A|B)P(B)}{P(A)}$$

Now we have expressed the probability of $B$ given $A$ in terms of the probability of $A$ given $B$. This is often useful, as in many problems the information on the probabilities may be easy to get one way round, but needed the other way round. The following card trick shows just such an example.

## ● *Example 13*

I have three cards. One, $C_1$, is black on both sides, the second, $C_2$, is red on both sides and the third, $C_3$, is red on one side and black on the other. Suppose I choose a card at random, and show you that one side of the card is red. What is the probability of the other side being red? Before you look at the solution have a guess.

SOLUTION
Our sample space $S$ has six events: three cards $C_1, C_2, C_3$ in conjunction with two sides. So the probability of each outcome is 1/6. Now, if I am only interested in which card I have picked, I can group the outcomes into three pairs, corresponding to the events $C_1, C_2$ and $C_3$, to get

$$P(C_1) = P(C_2) = P(C_3) = 1/3$$

Also, there are three red sides and three black sides, so I can also define two events $R$ and $B$, corresponding to the colour which my audience sees, and I have

$$P(R) = P(B) = 1/2$$

The probability we require is $P(C_2|R)$. It is not immediately obvious what this should be. However, the conditional probabilities the other way round *are* obvious, and in particular, $P(R|C_2)$ is 1 since card 2 has both sides red. Hence

$$P(C_2|R) = \frac{P(R|C_2)P(C_2)}{P(R)} = \frac{1 \times 1/3}{1/2} = 2/3$$

### Independence

### • *Example 14*

Consider the following pairs of events:

(a)  'It is raining in Blackpool' and 'Today is Thursday'
(b) . 'I throw a 6 on a die' and 'I flip a head on a coin'
(c)  'The first flip of a fair coin is a head' and 'The third flip is a head'
(d)  'Hospital admissions increase' and 'It is a cold winter'
(e)  'Wheat seeds grow better' and 'The wheat field was fertilised'

Some of these pairs of events are related and others are not. The fact that it is raining in Blackpool makes it no more likely that today is a Thursday than any other day of the week. Throwing a 6 and flipping a head are similarly unrelated. And it is reasonable to assume that when a coin is flipped, the results of previous flips have no effect—even if we had just had a run of 10 heads there is no reason to suppose that the next flip is more likely to be a tail rather than a head. In (d), however, there is a possible connection—the cold winter may be causing an increase in illness and hence an increase in hospital admissions, so we cannot assume that these events are unrelated. Similarly, we would expect the wheat's growth to be affected by the fertiliser, so it would not be reasonable to assume that the two events in (e) are unrelated.

This idea of events being unrelated is expressed in probability theory as *independence* of the two events. We say that two events $A$ and $B$ are **independent** if the occurrence of $B$ has no effect on the probability of $A$. So

$$P(A|B) = P(A)$$

If $A$ and $B$ are not independent then they are said to be dependent.

Hence in Example 14, (a), (b) and (c) are examples of independent events, whereas (d) and (e) are examples of dependent events.

### • *Example 15*

Let $A$ and $B$ be two events and $P(A) > 0$. If $B$ has no effect on $A$, is it possible for $A$ to have an effect on $B$?

SOLUTION
To answer this question we use Theorem 5, which states that

$$P(A \cap B) = P(A|B)P(B) = P(B|A)P(A)$$

together with

$$P(A|B) = P(A)$$

which states that $B$ has no effect on $A$. This gives

$$P(A)P(B) = P(B|A)P(A)$$

and since $P(A) > 0$ we can cancel it from both sides to show that $P(B|A) = P(B)$, so $A$ has no effect on $B$ either.

We have now shown that independence is a symmetric relation. If $A$ is independent of $B$, then $B$ is independent of $A$. One consequence of this is the following.

## ● *Theorem 6* ————————————————————

Events $A$ and $B$ are independent if and only if

$$P(A \cap B) = P(A)P(B)$$

Theorem 6 gives another way to establish whether two events are independent, and also shows the symmetry in the concept—$A$ is independent of $B$ if and only if $B$ is independent of $A$.

## ● *Example 16*

A card is selected at random from a pack. Let $A$ be the event that the card is below a 5 (i.e. 2, 3 or 4), and $B$ be the event that the card is a heart. Are $A$ and $B$ independent?

SOLUTION

$$P(A \cap B) = 3/52, P(A) = 12/52, P(B) = 13/52$$

Hence

$$P(A)P(B) = \frac{13 \times 12}{52 \times 52} = 3/52 = P(A \cap B)$$

Hence $A$ and $B$ are independent.

## ● *Example 17*

Suppose we flip a fair coin three times. Let event $A$ be 'we get three heads' and event $B$ be 'we get a head on the first flip'. Are $A$ and $B$ independent events?

SOLUTION

$$P(A \cap B) = P(A) = \tfrac{1}{2} \times \tfrac{1}{2} \times \tfrac{1}{2} = \tfrac{1}{8} \text{ and } P(B) = \tfrac{1}{2}$$

So $P(A)P(B) = \tfrac{1}{16} \neq P(A \cap B)$. Hence, $A$ and $B$ are dependent.

Theorem 6 also provides a simple way of calculating $P(A \cap B)$ if we know that $A$ and $B$ are independent. Returning to Example 14(c), let $A$ be 'head on first flip' and $B$ be 'head on third flip'. Then $P(A) = P(B) = \tfrac{1}{2}$, and if we assume that $A$ and $B$ are independent, then $P(A \cap B) = \tfrac{1}{2} \times \tfrac{1}{2} = \tfrac{1}{4}$.

## ● *Example 18*

In a class there are 4 left-handed men, 6 left-handed women and 6 right-handed men. How many right-handed women must be present if sex and handedness are to be independent when a student is selected at random?

SOLUTION

Let $A$ be the event that the student is a woman, $B$ the event that the student is left-handed and $x$ the number of right-handed women students in the class. So the number of women in the class is $6 + x$, the number of left-handed students is 10 and the total number of students is $16 + x$. Using

$$P(A)P(B) = P(A \cap B)$$

we see that we must have

$$\frac{6+x}{16+x} \times \frac{10}{16+x} = \frac{6}{16+x}$$

This is equivalent to

$$10(6 + x) = 6(16 + x)$$

and the solution is $x = 9$. The class must have 9 right-handed women.

Notice that in Example 18, the proportion of women amongst the left-handers is 6 out of 10, or 0.6, and with $x = 9$, the proportion of women amongst the right-handers is 9 out of 15, which is again 0.6. Since 'left-handed' and 'right-handed' are complementary events, this example illustrates a third way to see if two events are independent:

## ● *Theorem 7*

Events $A$ and $B$ are independent if and only if

$$P(A|B) = P(A|B^c)$$

### *An experiment*

It may or may not be the case that the proportions of left-handers are the same amongst men and women. Collect some information from your class-mates and investigate this question for yourself. Remember that the *observed* proportions amongst your class-mates can only be exactly equal by a fluke—just like in our coin-flipping experiments. So you will need a large class for the observed proportions to be reliable indications of the proportions in the population at large.

Left-handedness may confer an advantage in some two-person sports, because the unusual style of the left-hander gives them the advantage of surprising their opponent. Check the current list of top-rated tennis players, and find out what proportion are left-handers. Compare this with the proportion amongst your class-mates. Are left-handedness and tennis ability independent?

## ● *Example 19*

Let $A$ be the event that it rains in Blackpool tomorrow and $B$ be the event that tomorrow is a Thursday. Are $A$ and $B^c$ independent?

SOLUTION

Given that it rains tomorrow in Blackpool, there is no reason to suppose that tomorrow will not be a Thursday and we can show this as follows.

Firstly, express $A$ as the union of two mutually disjoint sets using the identity

$$A = (A \cap B) \cup (A \cap B^c)$$

Then, since this is a disjoint union we can use Kolmogorov's Axiom 3 to get

$$P(A) = P(A \cap B) + P(A \cap B^c) = P(A)P(B) + P(A \cap B^c)$$

where we have used the fact that $A$ and $B$ are independent. Hence

$$P(A \cap B^c) = P(A) - P(A)P(B) = P(A)\{1 - P(B)\}$$

and since

$$P(B^c) = 1 - P(B)$$

this gives

$$P(A \cap B^c) = P(A)P(B^c)$$

Hence, if $A$ and $B$ are independent then $A$ and $B^c$ are also independent.

## Bayes' theorem

### Example 20

A blood test is 95% effective in detecting a serious disease when it is in fact present. However, the test also gives a 'false positive' result for 1% of healthy people tested. It is known that 0.4% of the population have the disease. Suppose that you go for a test and the result is positive. Should you panic?

SOLUTION

Let $A$ be the event that the test is positive and $D$ the event that the person has the disease. We want to know $P(D|A)$. We know $P(A|D) = 0.95$, $P(A|D^c) = 0.01$ and $P(D) = 0.004$. Now

$$P(D|A) = \frac{P(D \cap A)}{P(A)} = \frac{P(A|D)P(D)}{P(A)}$$

and, as in Example 19, we can write

$$P(A) = P(A \cap D) + P(A \cap D^c) = P(A|D)P(D) + P(A|D^c)P(D^c)$$

Thus

$$P(D|A) = \frac{P(A|D)P(D)}{P(A|D)P(D) + P(A|D^c)P(D^c)}$$

$$= \frac{(0.95)(0.004)}{(0.95)(0.004) + (0.01)(0.996)} = 0.276\,163$$

So, although the test seems to be very accurate it turns out that false positives are more likely than true positives because the disease is so rare.

The result

$$P(D|A) = \frac{P(A|D)P(D)}{P(A|D)P(D) + P(A|D^c)P(D^c)}$$

is very useful and can be generalised to give a theorem due to Thomas Bayes (1702–1761), a clergyman and amateur statistician who lived in Britain. Although Bayes was an amateur, and none of his works were published in his lifetime, he was elected to the Royal Society in 1742. His essay on what is now called Bayesian inference, which has had a profound effect on modern statistical thinking, seems to have been ignored by his contemporaries. Coincidentally, the Royal Statistical Society now has its headquarters very close to Bayes' grave, which is in Bunhill Fields, Islington.

In order to state and prove Bayes' theorem in its most general form we need to introduce one new idea. A **partition** of the sample space is a collection of disjoint events $D_1, D_2, \ldots$ whose union is $S$, the sample space. Then, for any event $A$, every element in $A$ must also be in exactly one of the events $D_i$, so

$$P(A) = \sum P(A \cap D_i) = \sum P(D_i)P(A|D_i)$$

## Example 21

Suppose you are about to enter a lift which is showing a red warning light. The light may have come on for one of three different reasons: (a) the lift is faulty, (b) the lift is fine but the light has stuck on, and (c) the lift is going slowly. From years of experience using the lift, you know that the probabilities of the light showing, given each of the three different situations of the lift, are 0.8, 0.1 and 0.3 for situations (a), (b) and (c) respectively. You also know that the probabilities of the different situations occurring are 0.02, 0.9 and 0.08, again for (a), (b) and (c) respectively—notice, incidentally, that these last three probabilities must sum to 1, whereas this is not a requirement for the three *conditional* probabilities given previously, since they were conditional on *different* situations. When you discover the red light on, you want to know what is the probability that the lift is actually faulty.

SOLUTION
Let $D_1$ be the event that the lift is faulty and $R$ be the event that the red light is on. We want to know $P(D_1|R)$, and we know that in general,

$$P(D_1|R) = \frac{P(R|D_1)P(D_1)}{P(R)}$$

We know $P(R|D_1)$ and $P(D_1)$. To determine $P(R)$, we can write $R$ as the union of all possible situations. Let $D_2$ be the event that the lift is fine and $D_3$ the event that the lift is going slowly. Then, $D_1$, $D_2$ and $D_3$ form a partition, since they represent the only possible situations for the lift and they are mutually disjoint. So,

$$R = (R \cap D_1) \cup (R \cap D_2) \cup (R \cap D_3)$$

and

$$\begin{aligned} P(R) &= P(R \cap D_1) + P(R \cap D_2) + P(R \cap D_3) \\ &= P(R|D_1)P(D_1) + P(R|D_2)P(D_2) + P(R|D_3)P(D_3) \end{aligned}$$

So,

$$P(D_1|R) = \frac{P(R|D_1)P(D_1)}{P(R|D_1)P(D_1) + P(R|D_2)P(D_2) + P(R|D_3)P(D_3)}$$

We can use the sigma notation $\sum$ to write the denominator more neatly, giving

$$P(D_1|R) = \frac{P(R|D_1)P(D_1)}{\sum_{i=1}^{3} P(R|D_i)P(D_i)}$$

Now, we are told that $P(R|D_1) = 0.8$ and $P(D_1) = 0.02$, from which we can evaluate the numerator of the above expression as $0.8 \times 0.02 = 0.016$. Similarly, $P(R|D_2)P(D_2) = 0.1 \times 0.9 = 0.09$ and $P(R|D_3)P(D_3) = 0.3 \times 0.08 = 0.024$, from which we can evaluate the denominator as $0.016 + 0.09 + 0.024 = 0.13$. Hence, $P(D_1|R) = 0.016/0.13 = 0.123$, approximately.

Work out for yourself the conditional probabilities $P(D_2|R)$ and $P(D_3|R)$. As in Example 20, the 'obvious' explanation for the red light, namely that the lift is faulty, turns out not to be the most likely because it corresponds to a relatively rare situation.

In general, if $D_1, D_2, \ldots, D_n$ form a partition of $S$, then we can write

$$P(R) = P(R \cap D_1) + P(R \cap D_2) + \ldots + P(R \cap D_n)$$

$$= \sum_{i=1}^{n} P(R \cap D_i)$$

$$= \sum_{i=1}^{n} P(R|D_i)P(D_i)$$

Also, suppose we want to know the probability of $D_j$ having occurred, given that $R$ has occurred. Then we have the following result.

## ● *Theorem 8*

**Bayes' Theorem.** Let $D_1, D_2, \ldots, D_n$ be a partition of the sample space $S$ and $R$ be any event. Then

$$P(D_j|R) = \frac{P(R|D_j)P(D_j)}{\sum_{i=1}^{n} P(R|D_i)P(D_i)}$$

## ● *Example 22*

An electronics engineer is designing a television controller with four major components $C_1, C_2, C_3, C_4$. The engineer can run a series of tests to see what is the probability of the controller failing given that a specified component is faulty. She also knows from the manufacturers of the components how often each component fails. When a customer reports a fault with the controller, the engineer wants to know which is the most likely component to have failed. Which component should the engineer test first in order to reduce her average time to repair a controller?

SOLUTION
Let $C_i$ be the event that component $i$ is faulty and let $F$ be the failure of the controller.

Suppose $P(C_1) = 0.002$, $P(C_2) = 0.003$, $P(C_3) = 0.003$ and $P(C_4) = 0.001$. Suppose also that $P(F|C_1) = 0.4, P(F|C_2) = 0.2, P(F|C_3) = 0.1$ and $P(F|C_4) = 0.3$.

We want the conditional probabilities

$$P(C_j|F) = \frac{P(F|C_j)P(C_j)}{\sum_{i=1}^{4} P(F|C_i)P(C_i)}$$

Now,

$$\sum_{i=1}^{4} P(F|C_i)P(C_i) = (0.4 \times 0.002) + (0.2 \times 0.003) + (0.1 \times 0.003)$$
$$+ (0.3 \times 0.001)$$
$$= 0.002$$

and this gives

$$P(C_1|F) = \frac{P(F|C_1)P(C_1)}{\sum_{i=1}^{4} P(F|C_i)P(C_i)} = \frac{(0.4)(0.002)}{0.002} = 0.4$$

Similarly, $P(C_2|F) = 0.3$, $P(C_3|F) = 0.15$ and $P(C_4|F) = 0.15$.

This shows that the engineer will do best if she tests the components in the order $C_1, C_2$ and then either of $C_3$ or $C_4$.

## ● *Example 23*

In a bicycle factory, inner tubes are produced by three machines $M_1, M_2$ and $M_3$. Machine $M_1$ produces 50% of the tubes, $M_2$ produces 33% of the tubes and $M_3$ produces 17% of the tubes. Of the tubes made by $M_1$, 2% are defective, of tubes made by $M_2$, 5% are defective, and of tubes made by $M_3$, 3% are defective. A tube is selected at random and found to be defective. Which machine is most likely to have made it?

SOLUTION

Let $M_i$ be the event that the tube is made by machine $M_i$ and let $D$ be the event that the tube is defective. We want to know $P(M_i|D)$ for $i = 1, 2, 3$.

We know $P(M_i)$ and $P(D|M_i)$ for $i = 1, 2, 3$. So, using Bayes' theorem we have

$$P(M_j|D) = \frac{P(D|M_j)P(M_j)}{\sum_{i=1}^{3} P(D|M_i)P(M_i)}$$

Now,

$$\sum_{i=1}^{3} P(D|M_i)P(M_i) = (0.02 \times 0.5) + (0.05 \times 0.33) + (0.03 \times 0.17) = 0.0316$$

So,

$$P(M_1|D) = P(D|M_1)P(M_1)/0.0316 = (0.02 \times 0.50)/0.0316 = 0.316\,456$$

Similarly,

$$P(M_2|D) = (0.05 \times 0.330)/0.0316 = 0.522\,152$$

and

$$P(M_3|D) = (0.03 \times 0.170)/0.0316 = 0.161\,392$$

We obtain the largest value for $P(M_j|D)$ for $j = 2$.

## Example 24

In a mathematics department 70% of the students are from state schools and 30% are from private schools. State school pupils have a 2% failure rate and private school pupils a 4% failure rate in the first year exam. A student is picked at random from the list of fails. What is the probability that the student is from a state school?

SOLUTION

Let F denote the event 'fail', and $S$ the event 'state school'. We want $P(S|F)$ and are given $P(S) = 0.7$, $P(F|S) = 0.02$ and $P(F|S^c) = 0.04$. Then,

$$P(S|F) = \frac{P(F|S)P(S)}{P(F|S)P(S) + P(F|S^c)P(S^c)}$$

We calculate the denominator as

$$P(F|S)P(S) + P(F|S^c)P(S^c) = (0.02 \times 0.7) + (0.04 \times 0.3) = 0.026$$

So

$$P(S|F) = (0.02 \times 0.7)/0.026 = 0.538\,462$$

## Example 25

In a recent pre-election poll with just two candidates and everyone voting, 55% said they would vote Labour and 45% said they would vote Conservative. It is known that 90% of those who said they would vote Conservative actually voted Conservative, and 60% of declared Labour supporters actually voted Labour.

(a) What was the election result?
(b) What proportion of Conservative voters had said in the poll that they would vote Labour?

SOLUTION

Let $L$ be the event of actually voting Labour and $S$ the event of saying they would vote Labour. Then the information given can be expressed as the following three probabilities:

$$P(S) = 0.55, \quad P(L^c|S^c) = 0.9, \quad P(L|S) = 0.6$$

(a) Using the facts that $P(L|S^c) = 1 - P(L^c|S^c) = 0.1$ and $P(S^c) = 1 - P(S) = 0.45$, we can calculate the proportion of Labour voters as

$$P(L) = P(L|S)P(S) + P(L|S^c)P(S^c)$$
$$= (0.6 \times 0.55) + (0.1 \times 0.45) = 0.375$$

Hence $P(L^c) = 1 - P(L) = 0.625$, and the election will be won by the Conservatives.

(b) We want the conditional probability $P(S|L^c)$. Using Bayes' theorem, and the fact that $P(L^c|S) = 1 - P(L|S) = 0.4$, we can calculate this as

$$P(S|L^c) = \frac{P(L^c|S)P(S)}{P(L^c)} = \frac{(0.4 \times 0.55)}{0.625} = 0.352$$

## EXERCISES ON 5.5

1.  A box contains 10 items of type I, of which 3 are defective, and 20 items of type II, of which 5 are defective. An item is chosen at random from the 30 items in the box. Let $A = \{$item is of type I$\}$ and $B = \{$item is defective$\}$. Find:
    (a) $P(A)$
    (b) $P(A^c)$
    (c) $P(B)$
    (d) $P(A \cap B)$
    (e) $P(A \cup B)$
    (f) $P(A|B)$
    (g) $P(B|A)$
    Are $A$ and $B$ mutually exclusive? Are $A$ and $B$ independent? Would $A$ and $B$ have been independent if 6 of the 20 items of type II had been defective? (Try to answer this last question intuitively before doing the calculation.)

2.  Two dice are tossed. If the sum of their face values is 7, what is the probability that one of the dice has face value 5?

3.  True or false?
    (a) A gambler watching a roulette wheel has seen a run of reds. The gambler should bet on black at the next opportunity.
    (b) A card player is dealt 5 cards, all of which are red. The dealer deals her one more card. The player should bet on this next card being black.

4.  Seeds of three varieties of lettuce are scattered over a field. Exactly 50% of the seeds are of variety $A$, 30% are of variety $B$ and 20% are of variety $C$. From previous trials it is known that 50% of variety $A$ will germinate, 60% of variety $B$ will germinate and 70% of variety $C$ will germinate. If a particular seed germinates, what is the probability that it is of variety $C$?

5.  A testing device is used to screen electrical components for faults. It is known that the device will correctly detect a faulty component with probability 0.95, and will incorrectly declare a good component faulty with probability 0.02. The overall proportion of defective components is 0.01. A component is selected at random, and tested. The testing device declares the component to be faulty. What is the probability that the component is, in fact, faulty?

6.  Three weighted dice have the following probabilities of showing a 6:

    die $A$: 0.4
    die $B$: 0.2
    die $C$: 0.3

    One of the dice is selected at random and thrown twice. If neither throw was a 6, what is the probability that die $B$ was thrown?

7.  In a certain community, 4% of the population have alcohol poisoning. Suppose that the community's doctor correctly diagnoses alcohol poisoning with probability 0.95, and incorrectly diagnoses alcohol poisoning for a healthy patient with probability 0.05. If a patient is diagnosed as having alcohol poisoning, what is the probability that they actually have the disease?

8. A network is composed of components of types A and B connected in series and/or in parallel. A series sub-network fails if any one of its components fails. A parallel sub-network fails if all of its components fail. A type A component fails with probability 0.1, and a type B component fails with probability 0.01. Components fail (or not) independently of one another. Find the failure probability for each of the following networks.

(a)

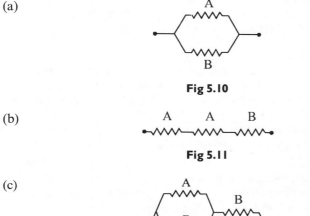

**Fig 5.10**

(b)

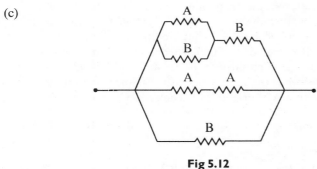

**Fig 5.11**

(c)

**Fig 5.12**

9. An experiment consists of flipping a coin a fixed number of times and recording the sequence of heads and tails. If different flips are independent, can you define an event which has probability $\frac{1}{3}$? How might you use the results from a sequence of coin flips to give a method of choosing amongst three possible actions with equal probability?

# 5.6 Projects

The following four projects all involve the idea of a **stochastic process**. This is any process which evolves according to the laws of probability. For example, if you look back at the graphs of the coin-flipping experiments in §5.2, the proportion $p_n$ of heads obtained from the first $n$ flips is a stochastic process.

It is possible to study the theory of stochastic processes in great detail, to obtain very general and powerful theorems about the behaviour of such processes. For each of the projects described here, all of the probability theory you need has been covered in the chapter. However, you will also have to think laterally, and to use your general mathematical knowledge, in order to solve each problem. You will learn more from the projects if you try to guess at least some of the answers before you work them out properly. You may also find it useful to carry out some experiments of your own, before attempting a theoretical solution.

## A simple gambling game

Two friends, Arthur and Betty, are playing a game to the following rules. Each time a fair coin is flipped, if it lands heads Arthur gives Betty 1p, and if it lands tails Betty gives Arthur 1p. The game ends when either Arthur or Betty has no money left.

1. Suppose that Arthur and Betty each have 2p at the start of the game. What is the probability that Arthur will win? Would your answer be the same if they each had 3p to start with? Or 5p to start with? Or an arbitrary amount $n$ pence to start with?
2. Suppose that at the start of the game, Arthur has 2p and Betty has 3p. Who is more likely to win? Can you generalise this result?
3. In general, suppose that Arthur has $x$ pence to start with, and Betty $(n - x)$ pence. Let $p_n(x)$ represent the probability that Arthur wins the game. Can you find an equation which relates $p_n(x)$ to $p_n(x - 1)$ and $p_n(x + 1)$? Can you solve this equation? You will need to say what are the values of $p_n(0)$ and $p_n(n)$.
4. Can you repeat the previous exercise when the coin is loaded, so that $P(H) = p$, a value not necessarily equal to 0.5? Given values for $p$ and $n$, what is a fair division of the money between Arthur and Betty (i.e. what is a fair value of $x$), so that Arthur and Betty each have an even chance of winning the game?

## The game of squash

In the game of squash, if the server wins a rally they add one point to their score and retain service for the next rally. If the receiver wins the rally, service is exchanged for the next rally but there is no change to the score. If the score reaches 8–8, the receiver may choose to play *first to 9 wins* or *first to 10 wins*. What choice should the receiver make? Does the choice depend on the relative skill of the two players, or their relative strengths as servers and receivers? A reasonable mathematical model for the game might be the following:

- the results of successive rallies are independent events
- when player $A$ is serving, they win the rally with probability $p_A$
- when player $B$ is serving, they win the rally with probability $p_B$

As in the previous project, it is a good idea to work out some simple special cases first, then see if you can generalise the method you used to solve these special cases.

1. Assume that $p_A = p_B = 0.5$. Write down all possible sequences of rallies which will result in player A winning 9–8, given that they are serving at 8–8. Deduce the probability that $A$ will win 9–8, given that $B$ chooses *first to 9 wins*.
2. Do the same for $A$ to win 10–8, and for $A$ to win 10–9, given that $B$ chooses *first to 10 wins*.

3. If the players are evenly matched, then we should have $p_A = p_B = p$, say, but there is no reason to assume that $p = 0.5$, because the server may be systematically advantaged or disadvantaged. Can you repeat the calculations from the first two exercises, but with an arbitrary value of $p$?

4. Use the result of Exercise 3 to decide what the receiver, B say, should choose, given that the two players are evenly matched. Does the answer depend on the value of $p$?

5. Finally, can you extend the calculation to the most general case of all, with $p_A \neq p_B$? Your final answer can be presented as a diagram of a unit square, corresponding to $0 \leq p_A \leq 1$ and $0 \leq p_B \leq 1$, with part of the square shaded to indicate the values of $p_A$ and $p_B$ for which the receiver, $B$, should choose *first to 9 wins*. Can you give a sensible, albeit approximate description of this shaded region in terms of the players' relative skills, and relative strengths of service?

## Rainfall in Tel Aviv

Extensive observations suggest that in Tel Aviv, Israel, the sequence of wet and dry days during the months of December, January and February can be described by the following simple mathematical model (proposed by Gabriel and Neumann in the article, 'A Markov chain model for daily rainfall occurrence at Tel Aviv' in *Quarterly Journal of the Royal Meteorological Society*, volume 88, pages 90–95):

- The probability that tomorrow will be dry, given that today was dry, is $\alpha$, irrespective of the weather before today.
- The probability that tomorrow will be dry, given that today was wet, is $\beta$, irrespective of the weather before today.

Meteorological data suggest that $\alpha$ is approximately 0.750 and that $\beta$ is approximately 0.338.

1. Deduce the probabilities that the day after tomorrow will be dry, given that today is dry or wet.

2. If $M$ is the $2 \times 2$ matrix

$$M = \begin{bmatrix} 0.750 & 0.250 \\ 0.338 & 0.662 \end{bmatrix}$$

calculate the matrix product $M^2$. How does this matrix relate to your answers to question 1?

3. Can you generalise this result to obtain a method for calculating the probabilities of a wet day in 3, 4, or more generally $n$ days' time, given that today is dry or wet?

4. If the probability of rain today is $p$, what is the probability of rain tomorrow? Express your answer in terms of the matrix $M$.

5. If you are told that the probability of rain tomorrow is the same as the probability of rain today, $p$ say, what is the value of $p$?

## Pass the pigs

The game of 'Pass the pigs' is a variation on dice-throwing. Instead of a pair of dice, each player throws a pair of small, identical model pigs. The pigs are constructed so that they can land in any of the following positions, with the scores indicated:

- *sider*: on either side (0 points, but see below!)
- *trotter*: on all four feet (5 points)
- *razorback* : on their back (5 points)
- *jowler*: balanced on their nose and two forelegs (10 points)
- *leaning jowler*: balanced on their nose, one foreleg and one ear (20 points)

**Fig 5.13** Pass-the-pigs™ (Registered Trade Mark, MB Games, Milton Bradley Ltd)

Each player scores points according to how the pigs land, adding the scores for each pig but doubling the score if both pigs land in the same position. So, for example, one sider and one trotter scores 5 points, one trotter and one jowler scores 15, a pair of jowlers scores 40. After each throw, the player can choose to throw again, and add whatever points they get to their previous score, or 'pass the pigs' to the next player. The first player to reach 100 points wins.

The tricky bit is that each pig has a spot marked on one side, and the player gets 1 point for a pair of siders, but **only** if both pigs land on the same side, i.e. you must have two spots or no spots showing. If the two pigs land on different sides, the player has suffered a 'pig-out' and gets no points for that round, and the pigs are passed to the next player. Even worse, if the two pigs are touching each other, the player loses all their points for the whole game, not just their score in the current round. So, the question is: what is a good strategy for playing 'Pass the pigs'?

You could approach this in many different ways, but whatever you do will have to combine some experimental work (for which you will need to buy the game—but we do not get any commission, honestly!) and some probability theory. Here are some points to consider :

- What are the probabilities associated with each possible position of a pig?
- Are the positions of the two pigs independent events?
- When you start the game, you must obviously throw the pigs to have any chance of getting a positive score, but once you have a score, you risk losing it if you keep throwing—and as your score increases, you have more to lose!
- If your opponent has close to 100 points, they have a good chance of winning on their next turn, so you may be inclined to take more risks, whereas if you are well ahead, a cautious strategy may be wiser.

As always, do not try to find the final answer immediately. Build up your answer by learning about the game. Answer the simpler questions first. Imagine a one-person version of the game, in which your objective is to reach 100 points in as few turns as possible. Then consider the two-person game. More than two players makes life very complicated, as you have to take into account all of the other players' scores to decide on the best strategy. So you may not be able to prove that your strategy is best—but you could play a lot of games with some friends who are not mathematicians and see how often you win!

# *Summary*

We use the word **experiment** to describe any procedure which has more than one possible result, or **outcome**. Specified *sets of outcomes* are called **events**.

The set of all possible outcomes of an experiment is called the **sample space**, $S$. The elements of a sample space are called **elementary outcomes**, often abbreviated to **outcomes.**

An **event** is any subset of $S$.

Intuitively the **probability** of an event, $A$ say, corresponds to the proportion of times that $A$ would occur if we repeated the experiment indefinitely. More formally a **probability function** P(.) is a mapping from $S$ to $\mathbb{R}$ which satisfies Kolmogorov's Axioms.

Kolmogorov's Axioms:

*Axiom 1*: For any event $A$, $P(A) \geq 0$.
*Axiom 2*: $P(S) = 1$.
*Axiom 3*: If $A$ and $B$ are disjoint events, then $P(A \cup B) = P(A) + P(B)$.

We write $P(A|B)$ to denote the probability of $A$ given that $B$ has occurred. Provided $P(B) > 0$, it is always true that

$$P(A|B) = \frac{P(A \cap B)}{P(B)}$$

Two events $A$ and $B$ are **independent** if the occurrence of $B$ has no effect on the probability of $A$, that is $P(A|B) = P(A)$. If $A$ and $B$ are independent,

$$P(A \cap B) = P(A)P(B)$$

**Bayes' theorem.** If events $D_j : j - 1, \ldots, n$ form a partition of the sample space $S$, and $R$ is any event, then

$$P(D_j|R) = \frac{P(R|D_j)P(D_j)}{\sum_{i=1}^{n} P(R|D_i)P(D_i)}$$

# 6 • Graphs

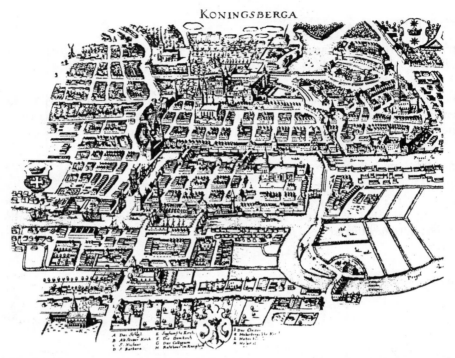

KONINGSBERGA

**Fig 6.1** The seven bridges of Königsberg (from *Graph Theory* 1736–1936, published by the Clarendon Press, Oxford)

In 1736 the people of Königsberg used to spend their Sunday afternoons trying to find a way of planning a walk to cross all seven bridges of their city exactly once and return to their starting point. This seemed to be impossible but no one could understand why. Leonhard Euler (pronounced Oiler) (1707–1783) was a prolific author (of 886 books and papers) and father (of 13 children). When presented with the Königsberg problem, he turned it into a question about the class of mathematical objects known as *graphs*, and we shall see his solution later.

Formally, a graph consists of a set of points with connections between some of the points. In this chapter, we first define some basic properties of graphs, and consider different ways of representing a graph algebraically. We then discuss various problems concerned with defining routes through a graph (of which the Königsberg problem is an example). Finally, we introduce you to the ideas of *graph colouring*, the most famous example of which is the *four-colour theorem*. This theorem states that four colours are sufficient to colour any map in such a way that no two adjacent countries have the same colour.

**Fig 6.2** Leonhard Euler (from Graph Theory 1736–1936, published by the Clarendon Press, Oxford)

# 6.1 Introduction

Since 1736 graphs have been used to describe many different situations. Figure 6.3 shows just a few examples: the map of the London Underground; friendship networks; printed circuit boards; and diagrams of the chemical bonds in molecules.

In the Underground map the graph shows the connections between different stations. We can see that some stations, for example Piccadilly Circus and Tottenham Court Road, are not connected directly, by which we mean that any route between them must visit other stations on the way. The graph does not show the actual geographical locations of the stations, or the distances between them—its purpose is only to demonstrate the interconnections between the stations.

Friendship networks are used by sociologists to study interpersonal relationships. In this case, the graph represents each person as a dot and joins two people

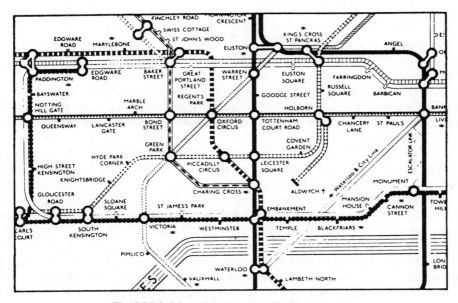

**Fig 6.3(a)** Map of the London Underground

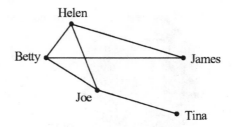

**Fig 6.3(b)** Friendship network

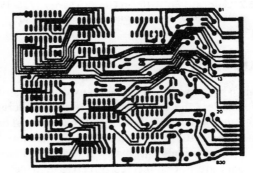

**Fig 6.3(c)** Printed circuit diagram

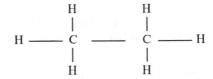

**Fig 6.3(d)** Chemical molecules

by a line if and only if they are friends. Similarly, Fig 6.3(b) shows a floor plan with rooms as dots joined by a line if and only if there is a door between them.

In a printed circuit board the lines of solder connect various electronic components. In this graph, it is important that none of the lines cross, otherwise there will be unwanted electrical connections at the crossing points. When the graphs are so complicated that crossings cannot be avoided, the circuits are divided up and put on different boards which are then sandwiched together. This then gives rise to two questions. Which circuits can be printed on one board with no crossings? And for circuits that cannot be printed on one board, what is the minimum number of boards needed? Problems of this kind are discussed in the section on planar graphs.

Finally, in the graph of a chemical molecule the dots represent the individual atoms in the molecule, whilst the lines represent the chemical bonds. The graph does not show how the atoms are spatially oriented, only how the different atoms are connected. Diagrams of this kind were introduced by Alexander Brown in 1864 to show that two molecules with the same chemical formula may have different chemical properties because of the different sets of connections between their atoms.

The common feature of all of these examples is that the graph consists of a set of points together with a set of lines, each line defining a connection between two points. The lengths of the lines and the positions of the lines and points on the page are assumed to be unimportant. In everyday language, the word 'graph' is used to describe many other kinds of diagram—for example, a chart of a hospital patient's temperature over a period of time or a 'pie-chart' of the proportions of votes won by different parties at an election.

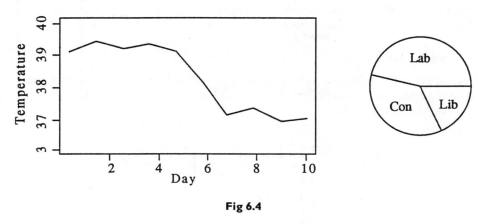

**Fig 6.4**

In this chapter, we will only use the word *graph* in its special sense.

## 6.2 Definitions and examples

The essential ingredients of any graph are a set of points, called **vertices**, and a set of lines, called **edges**, which form connections between pairs of points. We usually draw the points as solid dots to make them easier to see. The formal definition of a graph is the following.

A **graph** $G$ consists of a non-empty set $V(G)$ of **vertices** and a list $E(G)$ of unordered pairs of these elements called **edges**. If $v$ and $w$ are two vertices of $G$ then an edge of the form $vw$ is said to **join** $v$ and $w$.

### ● *Example 1*

For the graph $G$ in Fig 6.5, $V(G) = \{a, b, c, d\}$ and $E(G) = \{ab, ad, bd, bc\}$.

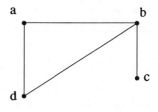

**Fig 6.5**

Two of the most basic properties of a graph are the number of vertices and the number of edges. We let $v(G)$ denote the number of vertices of a graph $G$, and $e(G)$ the number of edges of $G$. Two or more edges joining the same pair of vertices are called **multiple edges**. An edge joining a vertex to itself is called a **loop**. A graph with no loops or multiple edges is called a **simple graph**. A **subgraph** of a graph $G$ is a graph all of whose vertices belong to $V(G)$ and all of whose edges belong to $E(G)$.

It is often useful to know the number of edges at each vertex. The **degree** of a vertex $v$ of $G$ is the number of edges incident with $v$, and is denoted by $d(v)$. We use $\delta(G)$ and $\Delta(G)$ to denote the minimum and maximum degrees, respectively, of vertices of $G$. When evaluating these quantities, note that a loop adds two to the degree of the corresponding vertex.

## ● *Example 2*

The graph in Fig 6.6 has a pair of multiple edges from vertex $a$ to vertex $f$ and a loop at vertex $c$. The degrees are $d(a) = 3$, $d(b) = 3$, $d(c) = 4$ and $d(f) = 4$. Hence, $\delta(G) = 3$ and $\Delta(G) = 4$.

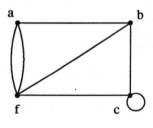

**Fig 6.6**

We can list the degrees in non-decreasing order to obtain the **degree sequence** for $G$. For the graph of Example 2 the degree sequence is $(3, 3, 4, 4)$.

## ● *Example 3*

Fig 6.7 shows a graph with degree sequence $(2, 3, 3, 4, 4, )$.

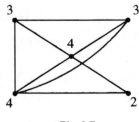

**Fig 6.7**

The degree sequence does not determine the graph, because different graphs may have the same degree sequence.

## • *Example 4*

Figure 6.8 shows another graph with degree sequence $(2, 3, 3, 4, 4)$.

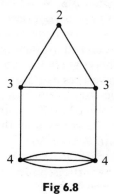

**Fig 6.8**

A graph where every vertex has the same degree is called **regular**. Figure 6.9 shows two examples of regular graphs.

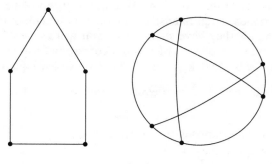

**Fig 6.9**

## • *Example 5*

Professor Harrison and her husband George are having a party at which there are four other couples. Some people at the party shake hands. Naturally, no couple shakes hands with each other. At the end of the party George comments that he and all the other guests have shaken hands a different number of times. The professor then tells George that in that case she can tell how many hands George shook. Since she was not watching him all the time, how does she know?

SOLUTION

We can use a graph to help us find the solution. We can draw a graph with ten vertices to represent the ten people at the party. Two vertices are joined by an edge if the two people have shaken each other's hand. The degree of each vertex then corresponds to the number of hands shaken by that person.

We know that the number of handshakes is different for George and the eight guests. What are the maximum and minimum number of handshakes possible for any one person? You may shake nobody's hand, so the minimum is 0 but what about the maximum? The maximum is 8. Why? Because although there are nine people excluding yourself, one of these is your partner. So we are looking for nine different numbers between 0 and 8, and of course there are only nine such numbers. So we know not only that George and all the other guests shook different numbers of hands, but also that these numbers are 0, 1, . . . , 8.

This allows us to label our ten vertices as *P* for the professor and 0, 1, . . . , 8 for George and the other guests, to correspond to the number of hands shaken by each of them. Which set of vertices is 8 joined to? There are only eight vertices available, as no vertices can be joined to the vertex 0 as this would imply that 0 shook somebody's hand, which is a contradiction. So who is 8's partner?

Now, which set of vertices is 7 joined to? There are only seven vertices available, as vertex 0 cannot be joined to any vertices, and vertex 1 can only be joined to one vertex but is already joined to vertex 8. So who is 7's partner? If you continue in this way you will see that the partners are 8 and 0, 7 and 1, 6 and 2, 5 and 3. Which leaves 4 as George.

We can use handshaking to help us derive one of the most basic results in graph theory. Consider the following situation. Amongst a group of people in a room some people shake hands. Each handshake involves two people. If you ask each person how many hands they have shaken the total will be twice the number of handshakes! If we make the vertices of a graph correspond to the people and the edges correspond to the handshakes, then we have the following result.

In any graph the number of edges is equal to half the sum of the degrees of the vertices.

Check this on some of the graphs in the examples above.

## ● *Theorem I* ──────────────────

**The Handshaking Lemma**.   In any graph the sum of all the vertex-degrees is equal to twice the number of edges.

PROOF
Every edge has two ends, so the sum of the degrees counts every edge twice.

From this result we can easily deduce the following corollaries.

## ● *Corollary I* ──────────────────

In any graph, the sum of all the vertex degrees is an even number.

## ● *Corollary 2* ──────────────────

In any graph, the number of vertices of odd degree is even.

## ● *Corollary 3* ──────────────────────

If $G$ is a regular graph with degree $r$ and $n$ vertices, then $G$ has $\frac{1}{2}nr$ edges.

We shall now look at some common examples of graphs.

- **Complete graphs**: these are simple graphs (Fig 6.10) in which each pair of distinct vertices is joined by an edge. The complete graph on $n$ vertices is denoted by $K_n$.

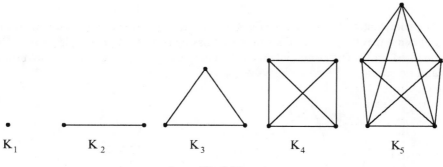

$K_1$      $K_2$      $K_3$      $K_4$      $K_5$

**Fig 6.10**

- **Bipartite graphs**: these are graphs (Fig 6.11) whose vertex set can be partitioned into two subsets $X$ and $Y$ so that each edge joins one vertex in $X$ to one in $Y$. Thus, in a bipartite graph there are no edges between the vertices of $X$, nor between the vertices of $Y$. A way of establishing whether a graph is bipartite is to try to colour each of the vertices either red or blue so that the only edges are between red and blue vertices.

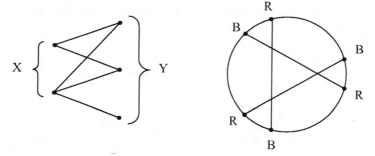

**Fig 6.11**

- **Complete bipartite graphs**: these are simple bipartite graphs (Fig 6.12) with bipartition $(X, Y)$ in which each vertex of $X$ is joined to every vertex of $Y$. The complete bipartite graph on $m$ vertices in $X$ and $n$ vertices in $Y$ is denoted by $K_{m,n}$.

$K_{2,3}$           $K_{3,3}$

**Fig 6.12**

- The **complement** of a graph $G$ (Fig 6.13) has the same set of vertices as $G$ but its edges are precisely those edges that are missing from $G$. The complement of $G$ is denoted by $\bar{G}$. The complement of the complete graph $K_n$ consists of $n$ vertices and no edges.

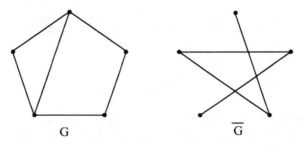

$G$           $\bar{G}$

**Fig 6.13**

- The **Petersen graph** (Fig 6.14) often occurs in graph theory examples—in fact, a whole book has been written about it, *The Petersen graph* by D. A. Holton and J. Sheehan, published by Cambridge University Press. The Petersen graph has 10 vertices, and is regular of degree 3. It was named after a Danish mathematician, Julius Petersen (1839–1910).

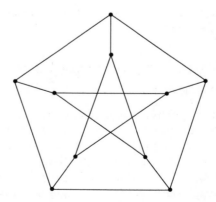

**Fig 6.14**

**1.** For the following degree sequences, either draw a graph with the given degree sequence or explain why there is no graph with such a sequence.
   (a) $(2, 2, 2)$
   (b) $(3, 3, 3, 3, 3)$
   (c) $(1, 2, 2, 3)$
   (d) $(2, 5, 5)$

**2.** How many edges has the complete graph on $n$ vertices?

**3.** True or false?
   (a) A simple graph may have loops but not multiple edges.
   (b) The graph $G$ shown in Fig 6.15 is a simple graph.
   (c) The graph $G$ shown in Fig 6.15 has degree sequence $(3, 3, 3, 4, 4)$.
   (d) A regular graph of degree 4 and $n$ vertices has $4n$ edges.
   (e) Bipartite graphs always have an even number of vertices.

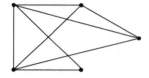

**Fig 6.15**

**4.** Draw $K_6$ and $K_{2,4}$.

**5.** What is the complement of the graph $K_{2,3}$? What is the complement of the graph $K_{m,n}$?

**6.** How many edges does the complement of a graph with $n$ vertices and $m$ edges have?

**7.** Let $G$ be a simple graph with at least two vertices. Show that $G$ contains two vertices of the same degree. (Remember the pigeonhole principle from Chapter 3.)

**8.** Show that there are no simple graphs with the following degree sequences:
   (a) $(7, 6, 5, 4, 3, 3, 2)$
   (b) $(6, 6, 5, 4, 3, 3, 1)$

**9.** Is the Petersen graph bipartite? Draw a bipartite graph on 10 vertices which is regular of degree 3.

# 6.3 Representations of graphs and graph isomorphism

In order to be able to store graphs in a computer we need to be able to represent the graphs in a form which the computer can understand and manipulate easily.

Here are two different ways using matrix representations.

## Adjacency matrix

Let $G$ be a graph with $n$ vertices labelled $v_1, v_2, \ldots, v_n$. The **adjacency matrix** $A(G)$ is the $n \times n$ matrix in which $A_{ij}$ is the number of edges joining $v_i$ and $v_j$.

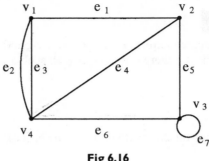

**Fig 6.16**

## Example 6

Here is the adjacency matrix for the graph shown in Fig 6.16:

$$\begin{pmatrix} 0 & 1 & 0 & 2 \\ 1 & 0 & 1 & 1 \\ 0 & 1 & 2 & 1 \\ 2 & 1 & 1 & 0 \end{pmatrix}$$

Note that the matrix is symmetric about the main diagonal. This will be true for all adjacency matrices, because whenever $v_i$ is joined to $v_j$, then by definition, $v_j$ will also be joined to $v_i$. If we label the vertices in a different order, then we will obtain a different adjacency matrix. However, the corresponding graphs will be essentially the same, since the labelling of the vertices is arbitrary.

### Incidence matrix

Let $G$ be a graph with $n$ vertices labelled $v_1, v_2, \ldots, v_n$ and $m$ edges labelled $e_1, e_2, \ldots, e_m$. The **incidence matrix** $I(G)$ is the $n \times m$ matrix in which $I_{ij}$ is the number of times that $v_i$ and $e_j$ are incident.

## Example 7

Here is the incidence matrix for the graph shown in Fig 6.16:

$$\begin{pmatrix} 1 & 1 & 1 & 0 & 0 & 0 & 0 \\ 1 & 0 & 0 & 1 & 1 & 0 & 0 \\ 0 & 0 & 0 & 0 & 1 & 1 & 2 \\ 0 & 1 & 1 & 1 & 0 & 1 & 0 \end{pmatrix}$$

Notice that the incidence matrix of a graph also depends on the particular ordering of the vertices and edges.

### Graph isomorphism

## Example 8

Figure 6.17 shows four pairs of graphs. For each pair, can you decide if the two graphs are the same or different?

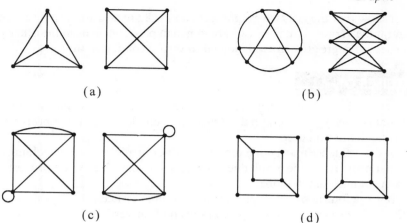

**Fig 6.17**

Graphs that are 'essentially the same' are called isomorphic. That is, although at first glance the two graphs may look different we can redraw them on the page to look the same. More formally we say that a graph $G$ is **isomorphic** to a graph $H$ if there exists a one-to-one mapping $\phi$ from $V(G)$ to $V(H)$ such that $\phi$ preserves adjacency, i.e. $uv \in E(G)$ if and only if $\phi(u)\phi(v) \in E(H)$. We then write $G \cong H$. Note that since $|V(G)| = |V(H)|$, any one-to-one mapping is equivalent to a relabelling of the vertices.

Thus, in order to check isomorphism we must find a labelling of each graph so that the correspondence is obvious. On the other hand, to show that two graphs are not isomorphic we would have to show that no labelling can produce the required agreement, and this can be quite difficult. It is often easier to show an obvious structural difference between the two graphs. We can easily check the number of vertices, number of edges and degree sequence of each graph. However, these all agree for the last pair of graphs in Example 8, yet the two graphs are not isomorphic. To see why, we need to look more closely at the structure of each graph.

## Example 9

Label the isomorphic pairs from Example 8 to show their isomorphisms.

SOLUTION
The solution is shown in Fig 6.18.

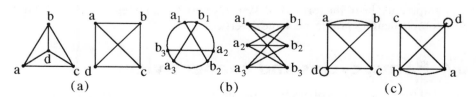

**Fig 6.18**

We can show that the last two graphs in Example 8 are not isomorphic because in one graph the vertices of degree two are joined in pairs, whereas in the other graph the vertices of degree two are only joined to vertices of degree three.

## EXERCISES ON 6.3

1. For the adjacency matrix in Example 6 work out all the row and column sums. What property of the graph does each row sum describe? What property does each column sum describe?

2. For the incidence matrix in Example 7, work out all the row and column sums. What property of the graph does each row sum describe? What property does each column sum describe?

3. Draw the four non-isomorphic graphs on three vertices.

4. Draw the eleven non-isomorphic graphs on four vertices.

5. Prove that the graphs shown in Fig 6.19 are not isomorphic.

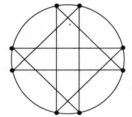

**Fig 6.19**

6. Show that the graphs shown in Fig 6.20 are isomorphic.

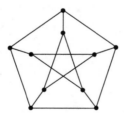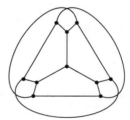

**Fig 6.20**

7. A simple graph which is isomorphic to its complement is called **self-complementary**. Show that the number of vertices of a self-complementary graph must be of the form $4k$ or $4k + 1$, where $k$ is an integer, and find all self-complementary graphs on 4 and 5 vertices.

# 6.4 Paths, cycles and connectivity

Many of the applications of graph theory involve finding a path from one vertex to another. For example, a major concern for the telephone companies is to decide the best route for a telephone call from one subscriber to another. Another example, faced by travelling sales representatives, is to plan their route each week so as to visit all their customers and return back to their home. The problem of finding the shortest possible route is known as the *Travelling Salesperson Problem*. It has been the subject of numerous papers, and in general there is no quick solution. By this, we mean that all general solutions involve explicit consideration of all possible routes to find the shortest—always possible in theory, since the number of possible routes is finite, but hopelessly impractical when the number of customers is large.

## ● *Example 10*

In the graph shown in Fig 6.21, find three different routes from *a* to *b*.

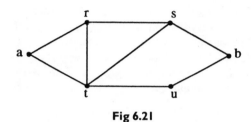

**Fig 6.21**

SOLUTION
1. *artsrtub*—in this route we repeat the edge *rt*.
2. *atrstub*—in this route we repeat the vertex *t*, but no edges are repeated.
3. *atsb*—in this route we do not repeat any edges or vertices.

It may be that the shortest possible route involves repeating some edges. However, in some applications we may want to disallow such routes. For example, a rambler planning a walk which takes in a specified set of mountain peaks may prefer not to retrace his or her steps along any paths which they had already walked, but would like to use the shortest possible route subject to this constraint. We shall now introduce some terminology so that we can distinguish between these different sorts of route.

A **walk** of length $k$ is a succession of $k$ edges of $G$ of the form

$$ab, bc, cd, df, fg, gh, \ldots, rs$$

denoted by $abc \ldots rs$. A walk of length $k$ all of whose edges are different is called a **trail** of length $k$. A walk of length $k$ all of whose edges *and* vertices are different is called a **path** of length $k$. We also have special terms if these begin and end at the same vertex. A **closed walk** is a succession of edges of the form

$$ab, bc, cd, \ldots, ef, fa$$

A closed walk all of whose edges are different is a **closed trail**. A closed walk all of whose vertices are different (except the first and last) is a **cycle**.

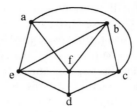

**Fig 6.22**

## Example 11

We can give examples of these in Fig 6.22. A walk is given by *acbedcaf*, a trail by *acbfae* and a path by *abcde*. A closed trail is *acbfeba* and a cycle is *dcfbed*.

So far, all our graphs have been in one piece, in the sense that you can visit all of the vertices by travelling along edges. It is usually easy to see if a graph is *connected* in this sense, but for some very complicated drawings of graphs it can be hard to see if there are some bits of the graph not connected to the rest.

A graph is **connected** if there is a path in *G* between any pair of vertices, and **disconnected** otherwise. Every disconnected graph can be split up into a number of connected subgraphs called **components**.

## Example 12

Decide which of the graphs in Fig 6.23 are connected.

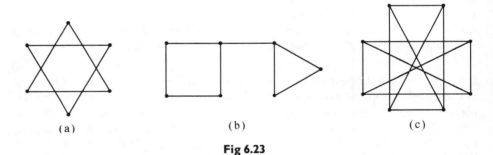

(a)  (b)  (c)

**Fig 6.23**

SOLUTION
Graphs (a) and (c) are disconnected and each has two components. Graph (b) is connected.

---

### EXERCISES ON 6.4

1.  In Fig 6.24 find:
    (a) a walk of length 10
    (b) a trail of length 6
    (c) a path of length 4
    (d) the number of different paths from *a* to *e*
    (e) cycles of length 1, 2, 3, 4 and 5.

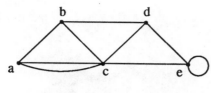

**Fig 6.24**

2. Find the length of the longest cycle in (a) $K_{3,3}$, (b) $K_7$, (c) $K_{2,3}$, (d) the Petersen graph.
3. Find the length of the shortest cycle in (a) $K_{3,3}$, (b) $K_7$, (c) $K_{2,3}$, (d) the Petersen graph.
4. Show that if a graph $G$ with $n$ vertices has minimum degree two then the graph has a cycle.
5. For the graph shown in Fig 6.25 obtain its adjacency matrix $A$. Work out $A^2$ and $A^3$. How many walks of lengths two and three are there from vertex $a$ to vertex $d$? How do these numbers relate to the entries $A^2(a, d)$ and $A^3(a, d)$?

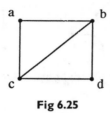

**Fig 6.25**

6. Prove that a graph and its complement cannot both be disconnected.
7. How many components has (a) $K_{2,3}$, (b) $\bar{K}_{2,3}$, (c) $\bar{K}_5$?

# 6.5 Trees

The graph of the chemical molecule shown in §6.1 (Fig 6.3(d)) has no cycles, and is an example of a **tree**. Arthur Cayley (1821–1895), a British mathematician who wrote almost 1000 papers, used trees in the 1870s to help him enumerate the numbers of chemical molecules.

Trees are also used in many applications. In linguistics, they are used to study sentence structure. In computer science, information is often stored in the form of a tree so that it can be easily accessed. In medicine, decision trees are used to help the process of diagnosis from a complicated pattern of symptoms and test results.

All the graphs shown in Fig 6.27 are examples of trees. Notice that trees do not have any cycles.

● *Example 13*

For each of the trees in Fig 6.27 count the number of edges and the number of vertices.

**Fig 6.26** Arthur Cayley (US Library of Congress)

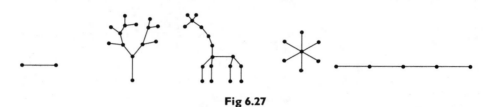

**Fig 6.27**

SOLUTION
Each tree $T$ has one fewer edge than the number of vertices.

There are many different ways to define a tree. We shall say that a **tree** is a connected graph with no cycles. As we have seen in the above example, it appears that any tree on $n$ vertices will have $n - 1$ edges.

## ● *Theorem 2*

Let $T$ be a tree on $n$ vertices. Then $T$ has $n - 1$ edges.

PROOF
We shall use induction on $n$, the number of vertices of $T$.

*Basis of induction:*
When $n = 1$ then $T$ is the graph $K_1$, which has no edges. Hence, $e = 0 = n - 1$.

*Inductive step:*
Now let us suppose the theorem is true for all trees with fewer than $n$ vertices. Let $T$ be a tree on $n \geq 2$ vertices. Let $uv$ be an edge of $T$. Then if we remove the edge $uv$ from $T$ we obtain the graph $T - uv$. We know that the graph we obtain is disconnected since $T - uv$ contains no $(u, v)$ path. If there were a path, say $u, v_1, v_2, \ldots, v$, from $u$ to $v$ then when we added back the edge $uv$ there would be a cycle $u, v_1, v_2, \ldots, v, u$ in $T$.

Thus, $T - uv$ is disconnected. The removal of an edge from a graph can disconnect the graph into at most two components. So $T - uv$ has two components. Call the two components $T_1$ and $T_2$. Then, both components are connected and are without cycles. Thus, $T_1$ and $T_2$ are trees and each has fewer than $n$ vertices. This means that we can apply the induction hypothesis to $T_1$ and $T_2$, to give

$$e(T_1) = v(T_1) - 1$$
$$e(T_2) = v(T_2) - 1$$

But the construction of $T_1$ and $T_2$ by removal of a single edge from $T$ tells us that

$$e(T) = e(T_1) + e(T_2) + 1$$

and that

$$v(T) = v(T_1) + v(T_2)$$

It follows that

$$e(T) = v(T_1) - 1 + v(T_2) - 1 + 1 = v(T) - 1$$

The theorem follows by the principle of mathematical induction.

A graph all of whose components are trees is called a **forest**. In a telephone network we do not want every subscriber to be connected directly to every other, as this would be very expensive. The cheapest network would be one in which the graph is connected but with no cycles. This will be a subgraph of our original graph called a spanning tree. A **spanning tree** is a subgraph of $G$ that includes every vertex of $G$ and is a tree.

## ● *Example 14*

For the graph shown in Fig 6.28 of all possible direct telephone connections between five cities in Australia, find two different spanning trees.

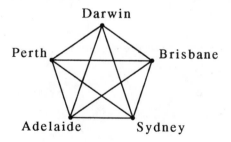

Darwin

Perth            Brisbane

Adelaide      Sydney

**Fig 6.28**

SOLUTION
The solution is shown in Fig 6.29.

A spanning tree of the telephone network will certainly cut out any unnecessary cycles in the graph but each connection will have a different cost associated with it, depending on the distance between the two vertices being joined. If our network has costs associated with each edge then we may need to find a spanning tree with

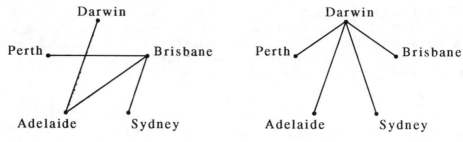

**Fig 6.29**

minimum total cost. If each edge of the graph has a length (or cost) associated with it then a **minimum spanning tree** is a spanning tree of minimum total length.

## ● *Example 15*

The following table shows the distances, in units of 100 km, between five cities in Australia. Find a minimum spanning tree.

|          | Adelaide | Brisbane | Darwin | Perth | Sydney |
|----------|----------|----------|--------|-------|--------|
| Adelaide | —        | 21       | 32     | 28    | 14     |
| Brisbane | 21       | —        | 35     | 44    | 10     |
| Darwin   | 32       | 35       | —      | 43    | 41     |
| Perth    | 28       | 44       | 43     | —     | 40     |
| Sydney   | 14       | 10       | 41     | 40    | —      |

SOLUTION
After some experimenting the minimum spanning tree is found to be as drawn in Fig 6.30.

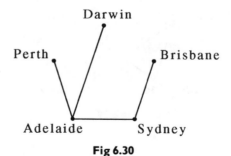

**Fig 6.30**

For bigger networks we will not be able to find the minimum so easily, and an algorithm is necessary. There is a simple algorithm, developed by Joseph Kruskal in the late 1950s, which can be used to find the minimum length spanning tree.

## Kruskal's algorithm

To find a minimum spanning tree, choose the shortest edge available each time, with the proviso only that we never choose an edge that makes a cycle.

This algorithm is an example of what is known as a *greedy algorithm* because at each stage in the process the best current choice is taken and we never have to worry about how that choice will affect future choices. In many non-greedy algorithms, the best choice now may lead to worse choices later and some *backtracking* is needed.

### ● Example 16

For the following table of distances, in miles, between six villages apply Kruskal's algorithm to find a minimum spanning tree:

|   | A | B | C | D | E | F |
|---|---|---|---|---|---|---|
| A | — | 5 | 6 | 12 | 4 | 7 |
| B | 5 | — | 11 | 3 | 2 | 5 |
| C | 6 | 11 | — | 8 | 6 | 6 |
| D | 12 | 3 | 8 | — | 7 | 9 |
| E | 4 | 2 | 6 | 7 | — | 8 |
| F | 7 | 5 | 6 | 9 | 8 | — |

SOLUTION

The smallest length is 2 between $B$ and $E$. The next smallest length is 3 between $B$ and $D$. The next smallest length is 4 between $A$ and $E$. The next smallest length is either 5 between $A$ and $B$ or 5 between $B$ and $F$. We cannot choose the edge between $A$ and $B$ as this would create the cycle ABE. But we can choose the edge between $B$ and $F$. There are three next shortest lengths of 6, between $C$ and $E$, $C$ and $F$, and $A$ and $C$. We can choose any one of these as none of them will create a cycle with the edges already chosen. Suppose we choose the edge $AC$.

We have found a minimum spanning tree of total length 20 miles: see Fig 6.31.

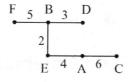

**Fig 6.31**

### ● Example 17

Apply Kruskal's algorithm to the Australian cities in Example 15.

SOLUTION

The shortest distance is 10 between Brisbane and Sydney. The next shortest is 14 between Adelaide and Sydney. The next shortest is 21 from Adelaide to Brisbane, but we cannot choose this edge as it would create a cycle. The next shortest is 28 from Adelaide to Perth. Finally, the next shortest is 32 from Adelaide to Darwin. We have found a minimum spanning tree of length 8400 km connecting the five cities.

## EXERCISES ON 6.5

1. True or False?
   (a) A tree has no cycles.
   (b) A tree is any graph with no cycles.
   (c) Every graph with $n$ vertices and $n - 1$ edges is a tree.
   (d) Every graph where any two vertices are connected by exactly one path is a tree.
   (e) Every tree is a bipartite graph.
2. Draw the six non-isomorphic trees on six vertices.
3. Prove that every tree has at least two vertices of degree one.
4. Draw a forest with eight vertices and five edges.
5. Prove that a forest with $k$ components has $n - k$ edges.
6. A saturated hydrocarbon is a molecule $C_mH_n$ in which every carbon atom has four bonds, every hydrogen atom has one bond and no sequence of bonds forms a cycle. Show that for every positive integer $m$, $C_mH_n$ can exist only if $n = 2m + 2$.
7. The following table shows the distances, in units of 10 km, between six towns in Britain. Find a minimum spanning tree.

|  | Aberdeen | Bristol | Cardiff | Dundee | Exeter | Fishguard |
|---|---|---|---|---|---|---|
| Aberdeen | — | 85 | 88 | 11 | 97 | 86 |
| Bristol | 85 | — | 7 | 74 | 14 | 25 |
| Cardiff | 88 | 7 | — | 77 | 20 | 18 |
| Dundee | 11 | 74 | 77 | — | 86 | 76 |
| Exeter | 97 | 14 | 20 | 86 | — | 37 |
| Fishguard | 86 | 25 | 18 | 76 | 37 | — |

# 6.6 Hamiltonian and Eulerian graphs

## Example 18

A historian, Dr Holmes, and a geographer, Dr Green, are on holiday on an island. Dr Holmes wishes to visit all the towns and Dr Green wishes to travel along every road. The graph shown in Fig 6.32 shows the towns (vertices) and roads (edges) on the island. Can you find a suitable route for each of them?

SOLUTION

Dr Holmes could use several routes which visit every vertex exactly once and return to the start, for example *abcdfea*.

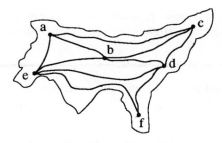

**Fig 6.32**

Dr Green has more of a problem. There does not seem to be any route that visits every road exactly once and returns to the start. We shall see in a moment why no route can be found.

The route for Dr Holmes is a cycle that visits every vertex of the graph exactly once. Such a cycle is called a **Hamiltonian cycle**. The route required for Dr Green is a closed trail that visits every edge of the graph exactly once. Such a trail is called an **Eulerian trail**. To see why there is not an Eulerian trail for Dr Green we observe that every time she enters and leaves a town on the trail this uses two of the edges at that vertex. Since she begins and ends at the same vertex this also uses two edges, hence every vertex must have even degree for such a trail to be possible. The graph representing the island does not have all vertices of even degree and hence there is no Eulerian trail.

We shall prove later in this section that as long as a graph has every vertex of even degree then there will always be an Eulerian trail.

## Eulerian graphs

We started this chapter with the problem put to Euler in 1736: to find a route for the people of Königsberg to cross all seven bridges in their town exactly once. We first turn the town plan into a graph. Let the vertices $A$, $B$, $C$ and $D$ be the four distinct land areas and join two vertices if there is a bridge between them (Fig 6.33).

Crossing a bridge corresponds to traversing an edge of the graph. So we are looking for a closed trail round the graph which uses every edge exactly once, an Eulerian trail. We say that a connected graph $G$ is **Eulerian** if there is an Eulerian trail. We say that a connected graph $G$ is **Eulerian** if there is an Eulerian trail.

Thus, to solve the Königsberg problem we want to know if the graph in Fig 6.33 is Eulerian. We have seen that a necessary condition for the existence of an

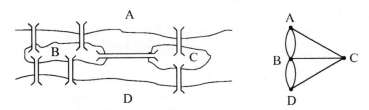

**Fig 6.33**

Eulerian trail is that the degrees of the vertices in the graph must all be even. We shall now prove that this condition is both necessary and sufficient.

## ● Theorem 3 ─────────────────────────────

A non-empty connected graph is Eulerian if and only if it has no vertices of odd degree.

PROOF

First of all we shall prove that if a non-empty connected graph is Eulerian then it has no vertices of odd degree.

Let $G$ be Eulerian. Then $G$ has an Eulerian trail which begins and ends at $u$, say. If we travel along the trail then each time we visit a vertex we use two edges, one in and one out. This is also true for the start vertex because we also end there. Since an Eulerian trail uses every edge once, the degree of each vertex must be a multiple of two and hence there are no vertices of odd degree.

Now we shall prove that if a non-empty connected graph has no vertices of odd degree then it is Eulerian. Let every vertex of $G$ have even degree. We will now use a proof by mathematical induction on $|E(G)|$, the number of edges of $G$.

*Basis of induction:*
Let $|E(G)| = 0$. Then $G$ is the graph $K_1$, and $G$ is Eulerian.

*Inductive step:*
Let $P(n)$ be the statement that all connected graphs on $n$ edges of even degree are Eulerian. Assume $P(n)$ is true for all $n < |E(G)|$. Since each vertex has degree at least two, $G$ contains a cycle $C$ (§6.4, Exercise 4).

Delete the edges of the cycle $C$ from $G$. The resulting graph, $G'$ say, may not be connected. However, each of its components will be connected, and will have fewer than $|E(G)|$ edges. Also, all vertices of each component will be of even degree, because the removal of the cycle either leaves the degree of a vertex unchanged, or reduces it by two. By the induction assumption, each component of $G'$ is therefore Eulerian.

To show that $G$ has an Eulerian trail, we start the trail at a vertex, $u$ say, of the cycle $C$ and traverse the cycle until we meet a vertex, $c_1$ say, of one of the components of $G'$. We then traverse that component's Eulerian trail, finally returning to the cycle $C$ at the same vertex, $c_1$. We then continue along the cycle $C$, traversing each component of $G'$ as it meets the cycle. Eventually, this process traverses all the edges of $G$ and arrives back at $u$, thus producing an Eulerian trail for $G$. This is shown in Fig 6.34.

Thus $G$ is Eulerian by the principle of mathematical induction.

## ● Example 19

Are the two graphs shown in Fig 6.35 Eulerian?

SOLUTION

The first graph has two vertices of odd degree, so is not Eulerian by Theorem 3. The second graph is Eulerian since every vertex has even degree. An Eulerian trail is *abcdeacebda*.

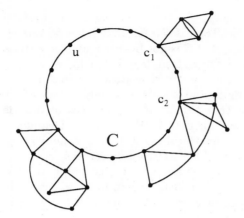

**Fig 6.34**

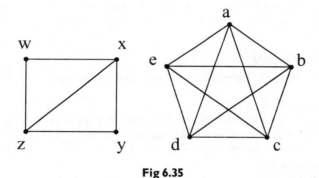

**Fig 6.35**

## ● *Example 20*

For the two graphs shown in Fig 6.35, can you trace their edges with a pencil without taking the pencil off the paper and without going over any edge twice?

### SOLUTION

The second graph is Eulerian, and we can therefore trace its edges as required, by tracing its Eulerian trail.

Although the first graph is not Eulerian, we can nevertheless trace its edges as required, by the following route: *zwxzyx*.

In the above example we required a trail that covered every edge exactly once, but we did not insist that the trail began and ended at the same place. A graph for which this can be done is called **edge-traceable.**

## ● *Corollary 4*

A connected graph is edge-traceable if and only if it has at most two vertices of odd degree.

PROOF

Let $G$ be edge-traceable. Then, if $G$ is Eulerian it has no vertices of odd degree and we are done. So assume that $G$ is not Eulerian, and let vertices $v$ and $w$ define the beginning and end of the open trail. If we add edge $vw$ we get an Eulerian trail in which, by Theorem 3, every vertex has even degree. If we now recover $G$ by removing $vw$, we see that $v$ and $w$ will be the only vertices of odd degree.

Now suppose that $G$ has at most two vertices of odd degree. If $G$ has no vertices of odd degree, then $G$ is Eulerian by Theorem 3 and so is edge-traceable. We know, from Corollary 2, that $G$ cannot have one vertex of odd degree. So, suppose that $G$ has exactly two vertices, say $v$ and $w$, of odd degree. By adding the edge $vw$ we obtain a connected graph all of whose vertices have even degree. By Theorem 3, we can find an Eulerian trail of this graph. Removal of edge $vw$ results in an open trail which includes every edge of $G$, so $G$ is edge-traceable.

Although Euler's Theorem tells us whether an Eulerian trail is possible it does not find one for us. To do this, we can use the following algorithm.

### Fleury's algorithm

If $G$ is an Eulerian graph, then the following steps produce an Eulerian tour:

*Step 1.*  Choose a starting vertex $u$.

*Step 2.*  Traverse any available edge, choosing an edge that will disconnect the remaining graph only if there is no alternative.

*Step 3.*  After traversing each edge, remove it (together with any vertices of degree 0 which result).

*Step 4.*  If no edges remain, stop. Otherwise, choose another available edge and go back to step 2.

## ● *Example 21*

Use Fleury's algorithm on the graph shown in Fig 6.36 to find an Eulerian trail.

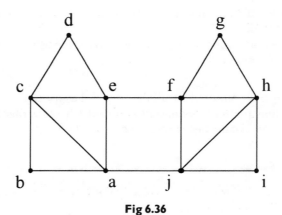

**Fig 6.36**

SOLUTION

We can start anywhere, so start at vertex $a$. We can choose $aj$, then $jf$, to define the start of our trail as $ajf$. At this stage we cannot choose the edge $fe$, since this would disconnect the graph and there are other alternatives. Proceeding in this way, we produce the partial trail $ajfghijhf$. Now we can choose $fe$, as there are no alternatives. An Eulerian trail is eventually given by

$ajfghijhfedcbacea$

## Hamiltonian graphs

A path that contains every vertex of $G$ exactly once is called a **Hamiltonian path** of $G$; similarly a **Hamiltonian cycle** of $G$ is a cycle that contains every vertex of $G$ exactly once.

**Fig 6.37** William Rowan Hamilton

The name comes from Sir William Rowan Hamilton (1805–1865), a leading mathematician of his time. He was appointed Astronomer Royal of Ireland when he was only 22. Hamilton also invented a game, called the Icosian game, which he sold to a Dublin toy manufacturer.

The Icosian game originally consisted of a wooden regular dodecahedron with the 20 corner points labelled with the names of cities: Brussels, Canton,..., Zanzibar. The object of the game was for the player to find a Hamiltonian cycle starting with any specified set of five initial vertices.

## Example 22

Figure 6.39 shows the graph of the Icosian game with the 20 vertices labelled.

If we start with the five initial vertices $PNMDF$, how many different Hamiltonian cycles are there?

SOLUTION

There are just two solutions with these five initial vertices, namely

$PNMDFKLTSRQZXWVJHGBCP$

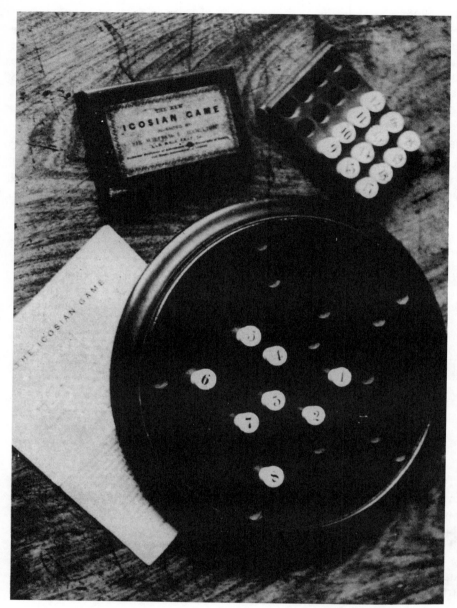

**Fig 6.38** Hamilton's Icosian Game (from *Graph Theory* 1736–1936, published by the Clarendon Press, Oxford)

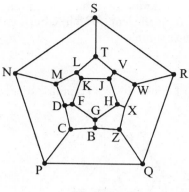

**Fig 6.39**

and

$$PNMDFGHXWVJKLTSRQZBCP$$

Hamilton was not the first to look for cycles that pass through every vertex—an older problem of this kind is the Knight's tour problem. This is described in the projects section at the end of this chapter.

## Example 23

Which of the graphs shown in Fig 6.40 are Hamiltonian?

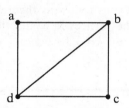 

**Fig 6.40**

SOLUTION

The first graph is Hamiltonian. A Hamiltonian cycle is *abcd*. The second graph is not Hamiltonian. We can see that this graph is bipartite, and that the two partitions have different numbers of vertices. Any cycle in a bipartite graph must have the same number of vertices in each partition, since any edge in a bipartite graph corresponds to a move from one partition to the other, hence this graph is not Hamiltonian.

We would like to have a necessary and sufficient condition for the existence of a Hamiltonian cycle, in the same way that Theorem 3 provides a necessary and sufficient condition for a graph to be Eulerian. However, this remains one of the main unsolved problems in graph theory. We shall show instead an example of a necessary condition and an example of a sufficient condition, but no one condition which is both necessary and sufficient. A simple but useful necessary condition is illustrated in the following example.

## ● *Example 24*

Does the graph shown in Fig 6.41 have a Hamiltonian cycle?

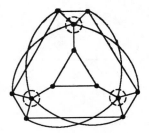

**Fig 6.41**

SOLUTION
This graph does not have a Hamiltonian cycle. If we remove the three circled vertices and their associated edges then the graph becomes disconnected into four pieces. If the original graph had a Hamiltonian cycle and we removed three vertices from the cycle, then the cycle would disconnect into at most three pieces. Since the graph has disconnected into more than three components it cannot have had a Hamiltonian cycle.

The idea in the above example is the essence of the following theorem. We use the notation $G - S$ to mean the graph $G$ with the vertices in $S$ and their associated edges removed. Also for any graph $G$ we use $k(G)$ to denote the number of components of $G$.

## ● *Theorem 4*

Let $G$ be any Hamiltonian graph. For every non-empty proper subset $S$ of $V(G)$,

$$k(G - S) \leq |S|$$

PROOF
Let $G$ be a Hamiltonian graph and $C$ be a Hamiltonian cycle of $G$. Let $S$ be a non-empty proper subset of $V(G)$. Then, if we remove the set of $S$ vertices from $C$, the cycle will fall into at most $|S|$ pieces. That is,

$$k(C - S) \leq |S|$$

However, $C - S$ is a subgraph that includes every vertex of $G - S$, and so $G - S$ cannot have more components than $C - S$. Thus,

$$k(G - S) \leq k(C - S)$$

Hence,

$$k(G - S) \leq |S|$$

We can use this necessary condition to show that certain graphs are not Hamiltonian. However, because the condition is not sufficient it is possible to find a non-Hamiltonian graph for which the removal of any set of vertices $S$ leaves no more components than the number of vertices in $S$.

### ● *Example 25*

Are the two graphs shown in Fig 6.42 Hamiltonian?

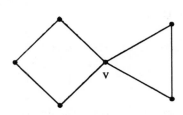
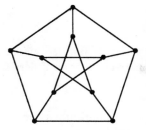

**Fig 6.42**

SOLUTION

In the first graph $G$, if we let $S = \{v\}$ then $G - S$ has two components and $|S| = 1$. Hence, by Theorem 4, $G$ is not Hamiltonian.

By trial and error we can see that the Petersen graph is not Hamiltonian. However, there is no set of vertices whose removal leaves a graph with more components than the number of vertices removed.

The next theorem, due to G. A. Dirac and published in 1952, is an example of a sufficient condition for a graph to be Hamiltonian.

### ● *Theorem 5* ─────────────

**Dirac's Theorem.**   Let $G$ be a simple graph with $v \geq 3$ vertices. If every vertex has degree at least $\frac{1}{2}v$ then $G$ is Hamiltonian.

A proof of this theorem can be found in *Introduction to Graph Theory*, by Robin J. Wilson, published by Longman.

### ● *Example 26*

Show that the two graphs shown in Fig 6.43 are Hamiltonian.

SOLUTION

The first graph has seven vertices and every vertex has degree at least four. So Dirac's Theorem tells us that this graph is Hamiltonian. The second graph is clearly Hamiltonian by inspection, but there are seven vertices and the degree of each vertex is only two. Hence Dirac's theorem is sufficient but not necessary.

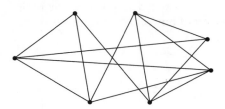

**Fig 6.43**

There are many puzzles whose solution relies on being able to find either a Hamiltonian cycle or an Eulerian trail:

## Puzzles

- How many continuous pen strokes are required to draw the diagram shown in Fig 6.44 without covering any line twice?

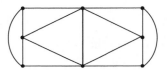

**Fig 6.44**

- Given a set of dominoes, can you arrange them in a ring so that the numbers on the adjacent ends of the dominoes coincide?
- Can you draw a continuous line that passes through every edge of the graph shown in Fig 6.45 exactly once?

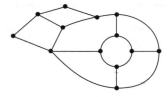

**Fig 6.45**

- A mouse eats its way through a $3 \times 3 \times 3$ block of cheese. It starts at one corner. Can it finish with the centre cube?

## EXERCISES ON 6.6

1. Decide whether the graphs shown in Fig 6.46 are:
   (i)   Eulerian but not Hamiltonian
   (ii)  Hamiltonian but not Eulerian

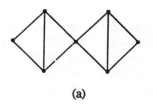 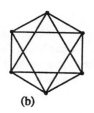  

(a)  (b)  (c)  (d)

**Fig 6.46**

    (iii)   Eulerian and Hamiltonian
    (iv)   neither Hamiltonian nor Eulerian
**2.**    True or false?
    (a)  All regular graphs of degree 4 are Eulerian.
    (b)  All connected regular graphs of degree 2 are Hamiltonian.
    (c)  If $G$ is Hamiltonian then $\delta \geq n/2$ (where $n$ is the number of vertices).
    (d)  If $G$ is not Hamiltonian then $\delta \leq n/2$.
    (e)  All bipartite graphs are Eulerian.
    (f)  All bipartite graphs with the same number of vertices in each partition are Hamiltonian.
**3.**    Find the values of $r$ and $s$ for which the graphs $K_{r,s}$ are Eulerian.
**4.**    A pavement sweeper must sweep the pavements on both sides of every street in a certain town. Is it possible for the sweeper to construct a route that covers every side of each street exactly once?
**5.**    Show that if a graph has $2t$ vertices of odd degree, then the minimum number of pen strokes to trace the graph is $t$.
**6.**    Find two Hamiltonian cycles in the Icosian game starting with *BGHJK*.
**7.**    Find the values of $r$ and $s$ for which the graphs $K_{r,s}$ are Hamiltonian.
**8.**    Find a cycle of length 9 in the Petersen graph. Show that the Petersen graph is not Hamiltonian.
**9.**    Find a regular, non-Hamiltonian graph with degree at least $12(v-1)$.

# 6.7 Planar graphs

A graph is said to be **planar** if it can be drawn in the plane in such a way that no two edges meet each other except at a vertex.

## ● *Example 27*

Which of the graphs shown in Fig 6.47 are planar?

 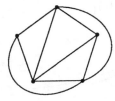 

**Fig 6.47**

SOLUTION

The second graph is clearly planar as it has been drawn so that none of its edges cross. However, the first graph is also planar because we can draw it so that none of its edges cross. The third graph does not look as if it is planar, but can you be sure there is not a redrawing which will have no edges crossing? Later on in this section we prove that this graph is not planar. Figure 6.48 shows a plane drawing of the first graph.

**Fig 6.48**

A particular graph may not look planar at first glance because the edges cross. However, as in Example 27 it may be possible to redraw the graph so that there are no crossings. A drawing of a planar graph without crossings is called a **plane** drawing. Here is a well-known children's puzzle.

**The Utilities Problem.**   There are three houses which are each to be connected to the three services: Gas, Water, Electricity. In order that the maintenance of any one pipe does not affect any of the others, the pipes must be laid so that they do not cross each other. Can you find a way to position each of the nine pipes so that none of the pipes cross?

   If we draw this puzzle as a graph (Fig 6.49), then it reduces to the question: is the graph $K_{3,3}$ planar? In order to answer this question, we need to find a property common to all planar graphs.

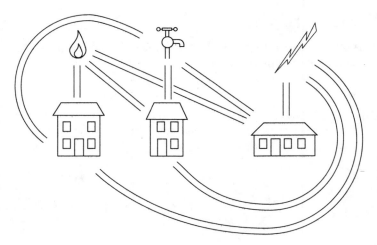

**Fig 6.49**

## Euler's formula

If $G$ is a planar graph, then any plane drawing of $G$ divides the plane up into regions called **faces**. The **degree of a face**, written $d(f)$, is the number of edges which surround the face. In counting the number of faces, we always consider the outside region to be a face.

## Example 28

For the planar graph shown in Fig 6.50 give the number of faces and the degree of each face.

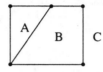

**Fig 6.50**

SOLUTION
There are three faces. Face $A$ has degree 3, $B$ has degree 4 and $C$ has degree 5. The sum of the degrees of the faces is 12, and there are 6 edges.

In the above example, the sum of all the degrees of the faces is equal to twice the number of edges in the graph. This will always be the case, because each edge borders two different faces and so will be counted twice in evaluating the sum of the degrees of the faces. This gives the result that for any planar graph,

$$\sum_{f_i \in F} d(f_i) = 2e$$

## Example 29

For the planar graph shown in Fig 6.51 count the number of vertices, edges and faces. Do not forget the outside face! What does $v - e + f$ equal? Draw another planar graph and see if you get the same result for $v - e + f$.

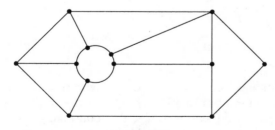

**Fig 6.51**

SOLUTION
We can see that $v = 12$, $e = 18$ and $f = 8$. So $v - e + f = 2$.

Our next theorem proves that for any planar graph, $v - e + f = 2$.

## ● *Theorem 6*

If $G$ is a connected planar graph with $v$ vertices, $e$ edges and $f$ faces, then

$$v - e + f = 2$$

PROOF
We shall use induction on $f$.

*Basis of induction:*
Let $f = 1$. Then, since $G$ is connected it must be a tree. Hence, by Theorem 2, $e = v - 1$. Hence, $v - e + f = v - (v - 1) + 1 = 2$, and the theorem holds.

*Inductive step:*
Let $P(f)$ be the statement that $v - e + f = 2$ for all graphs with $f$ faces.

Assume $P(f)$ is true for all $f$ less than $t$. Let $G$ be any connected planar graph with $v$ vertices, $e$ edges and $t \geq 2$ faces. Since $t \geq 2$, $G$ is not a tree and must therefore have at least one cycle. Choose any edge, $e_1$ say, of $G$ that is in a cycle. Then $G - e_1$ is a connected planar graph with $v$ vertices, $e - 1$ edges and $t - 1$ faces, since the two faces of $G$ separated by $e_1$ combine to form one face of $G - e_1$.

By the induction hypothesis applied to $G - e_1$,

$$v - (e - 1) + (t - 1) = 2$$

Hence,

$$v - e + t = 2$$

which establishes $P(t)$.

The theorem follows by the principle of mathematical induction.

We want to know if the graph for the Utilities Problem is planar. We know that for planar graphs, $v - e + f = 2$. For the graph $K_{3,3}$, we know that $v = 6$ and $e = 9$. How many faces are there? To count the number of faces directly, we need to produce a plane drawing of the graph—but this begs the question of whether the graph is, in fact, planar! To solve the problem, we need to find a formula which defines a relation on $v$ and $e$ for planar graphs which does not involve $f$.

## ● *Example 30*

Show that the graph $K_{3,3}$ is not planar.

SOLUTION
Since there are no loops, multiple edges or triangles (cycles of length three) the smallest cycle in the graph has length four. Thus if the graph $K_{3,3}$ is planar, then the degree of each face is at least 4. Hence the total number of edges around all the faces is at least $4f$. Since every edge borders two faces, the total number of edges around all the faces is $2e$, so we have established that $2e \geq 4f$, which is equivalent to

$$2f \leq e$$

If we combine this with Euler's formula, $v - e + f = 2$, we get

$$2f = 4 - 2v + 2e \leq e$$

which is equivalent to

$$e \leq 2v - 4$$

For the graph $K_{3,3}$, the number of vertices is 6 and so $2v - 4 = 8$. But the number of edges is 9, and so $e \not\leq 2v - 4$. Hence, the graph cannot be planar.

## ● *Example 31*

Show that the graph $K_5$ is not planar.

SOLUTION

As in the proof that $K_{3,3}$ is not planar, we consider the length of the smallest cycle in $K_5$. Since $K_5$ is a simple graph, the smallest possible length for any cycle of $K_5$ is three. We shall suppose that the graph is planar. Then, every face is bounded by at least three edges, so the total number of edges around all the faces must be at least $3f$. But we know that the total number of edges around all the faces is $2e$, which implies that

$$3f \leq 2e$$

If we combine this with Euler's formula, $v - e + f = 2$, we get

$$3f = 6 - 3v + 3e \leq 2e$$

which is equivalent to

$$e \leq 3v - 6$$

For the graph $K_5$, the number of vertices is 5 and so $3v - 6 = 9$. But the number of edges is 10, hence $e \not\leq 3v - 6$ and the graph is not planar.

We have now shown that the graphs $K_5$ and $K_{3,3}$ are not planar. These are not arbitrary choices of graph to show non-planarity. A very surprising result is that *any* graph that is not planar always contains (in a rather special way) one of these two graphs.

If we insert vertices on the edges of a graph as shown in Fig 6.52, we obtain a **subdivision** of the original graph.

Clearly, any subdivision of $K_5$ or $K_{3,3}$ is also non-planar. Less obviously, the converse is also true. This was proved in 1930 by the Polish mathematician Kazimierz Kuratowski (1896–1980).

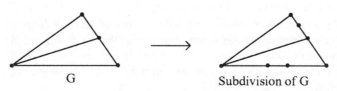

G            Subdivision of G

**Fig 6.52**

## ● *Theorem 7*

**Kuratowski's Theorem.** A graph is planar if and only if it does not contain a subdivision of $K_5$ or $K_{3,3}$.

For a proof of Kuratowski's Theorem see *Applications of Graph Theory* by J. A. Bondy and U. S. R. Murty, published by Macmillan.

## ● *Example 32*

We can use Theorem 7 to see that the Petersen graph is not planar because it contains a subdivision of $K_{3,3}$: see Fig 6.53.

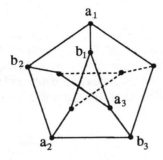 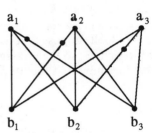

**Fig 6.53**

## EXERCISES ON 6.7

1. True or false?
   (a) A graph is planar if every drawing of it in the plane has no crossings.
   (b) $K_{3,3}$ is planar.
   (c) $K_{2,3}$ is planar.
   (d) $K_6$ is not planar.
   (e) $K_5$ can be drawn on a torus (doughnut) without crossings.
   (f) Every subgraph of a planar graph is planar.
   (g) Every subgraph of a non-planar graph is non-planar.

2. Let $G$ be a regular planar graph with degree 4 and eight vertices. How many faces does $G$ have?

3. What is the smallest length of cycle in the Petersen graph? Adapt the proof that $K_{3,3}$ is non-planar to show that the Petersen graph is non-planar.

4. The **dual** graph of $G$ is obtained by placing a vertex in every face of $G$ and an edge joining every two vertices in neighbouring faces of $G$. A planar graph $G$ has four vertices and eight faces. How many vertices and edges will the dual graph have?

5. Give an example of a planar graph where every vertex has degree 5.

6. Let $G$ be a connected planar graph. Show that the dual of $G$ is Eulerian if and only if $G$ is bipartite.

7. The **thickness** of a graph is the minimum number of planar graphs into which the graph can be decomposed. Find the thickness of
   (a) $K_{3,3}$

(b) $K_5$

(c) $K_7$.

# 6.8 Graph colouring

There are many different ways to colour a graph, each with different applications. We shall look at three different ways in this section.

## Edge colourings

A $k$-**edge colouring** of a (loopless) graph $G$ is an assignment of $k$ colours to the edges of $G$ in such a way that any two edges meeting at a common vertex have different colours. The **chromatic index** of $G$, denoted by $\chi'(G)$, is the smallest number $k$ for which $G$ has a $k$-edge colouring.

## ● Example 33

Colour the edges of each of the two graphs shown in Fig 6.54 with the minimum number of colours. What are their chromatic indices?

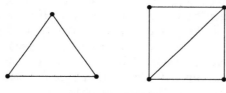

**Fig 6.54**

SOLUTION

The first graph has two edges at each vertex so we need at least two colours. But we can see that the third edge needs a new colour, hence the chromatic index for this graph is three. The other graph has at most three edges at each vertex, so we need at least three colours and in fact such a colouring is possible. Hence, the chromatic index is again three.

## ● Example 34

Figure 6.55 shows three bipartite graphs. We can see that for each graph the chromatic index is equal to the maximum degree, $\Delta$.

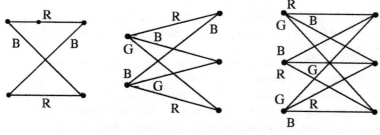

**Fig 6.55**

We can prove that for all bipartite graphs, $\chi'(G) = \Delta$. This was first proved by the Hungarian mathematician Dénes König (1884–1944). König wrote the first book on graph theory in 1936, *Theorie der endlichen und unendlichen Graphen.*

● *Theorem 8* ───────────────────────────────────

**König's Theorem.**   If $G$ is a bipartite graph whose maximum vertex degree is $\Delta$, then $\chi'(G) = \Delta$.

PROOF
We shall prove this by induction on $m$, the number of edges of $G$.

*Basis of induction:*
When $m = 1$, we have $\chi'(G) = 1$ and $\Delta = 1$.

*Inductive step:*
Let $P(n)$ be the proposition that all bipartite graphs $G$ with $n$ edges have $\chi'(G) = \Delta$. Assume that $P(n)$ holds for all $n < m$. Let $G$ be a bipartite graph with $m$ edges and maximum degree $\Delta$. Let $e = (v, w)$ be an edge of $G$. Then, $G - e$ has $m - 1$ edges and maximum degree at most $\Delta$, and is thus $\Delta$-colourable by the induction assumption. We now try to put a colour on the edge $e$. There will be at least one colour missing at both the vertices $v$ and $w$. If there is the same colour missing at both $v$ and $w$, then we can assign this colour to the edge $e$ and so complete the colouring of $G$. If this is not the case, suppose that blue is missing at $v$ and red is missing at $w$. Now, consider all the vertices that can be reached from $v$ by a path of red/blue edges. This path does not include $w$, since $G$ is bipartite and colours alternate along the path. We can interchange the colours on the red/blue path from $v$. Thus, red is now missing at both $v$ and $w$ and we can complete our colouring by colouring $e$ red.

There is an obvious lower bound for $\chi'(G)$. If $\Delta$ is the largest degree in $G$, then $\chi'(G) \geq \Delta$. The upper bound is surprising. It was proved by V. G. Vizing in 1964 that you only ever need at most one more colour than the maximum degree.

● *Theorem 9* ───────────────────────────────────

If $G$ is simple, then

$$\Delta \leq \chi'(G) \leq \Delta + 1$$

In the next example we shall find a way to establish whether or not a graph is $\Delta$-colourable.

● *Example 35*

Find $\chi'(K_5)$ and $\chi'(K_6)$.

SOLUTION
The graph $K_5$ has $\Delta = 4$. Since there are five vertices the maximum number of edges of any one colour is 2. Therefore with 4 colours we could colour at most 8 edges. However, there are 10 edges in $K_5$ and so we need at least 5 colours. A possible 5-colouring is shown in Fig 6.56. Note that parallel edges have the same

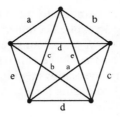

**Fig 6.56**

colour and that at each vertex there is a different colour missing, i.e. not used on any edge at that vertex.

In order to colour $K_6$, start with the colouring of $K_5$ shown in Fig 6.56. Now add a new vertex joined to the original 5 vertices by 5 edges, each coloured with the colour missing at the corresponding vertex. This gives a 5-colouring of $K_6$: see Fig 6.57.

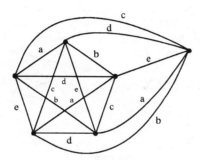

**Fig 6.57**

We can generalise the argument used in Example 35 to obtain the following theorem.

## ● *Theorem 10*

If $G$ is a complete graph, then

$$\chi'(K_{2n}) = 2n - 1$$

and

$$\chi'(K_{2n+1}) = 2n + 1$$

PROOF

The graph $K_{2n+1}$ has $\Delta = 2n$. There is no $2n$-edge colouring, since the maximum number of edges that can be assigned any one colour is $n$ and thus using $2n$ colours we could only colour $2n^2$ edges. But $K_{2n+1}$ has $2n^2 + n$ edges. Thus, by Theorem 9, $\chi'(K_{2n+1}) = 2n + 1$.

We can explicitly colour the edges of $K_{2n+1}$ by drawing the vertices in the form of a regular $(2n + 1)$-sided polygon and colouring the boundary edges in $2n + 1$ different colours. Each of the remaining edges is then coloured with the same colour as the boundary edge parallel to it. For $K_{2n}$, we can prove that $\chi'(K_{2n}) = \Delta = 2n - 1$ by giving an explicit colouring.

If $n > 1$, choose any vertex $v \in K_{2n}$ and consider $K_{2n} - v$. This will be $K_{2n-1}$, which we can $(2n - 1)$-colour by the above construction. At each vertex $w_i$, for $i = 1, \ldots, 2n - 1$, there is exactly one colour missing and this colour can be put on the edge $w_i v$. This gives a $2n - 1$ colouring for $K_n$.

## Vertex colourings

A **$k$-vertex colouring** of a loopless graph $G$ is an assignment of $k$ colours to the vertices of $G$ such that adjacent vertices receive different colours. The **chromatic number** of $G$, written $\chi(G)$, is the smallest number $k$ for which $G$ has a $k$-vertex colouring.

### ● *Example 36*

Find the minimum number of colours to colour the vertices of the graphs shown in Fig 6.58.

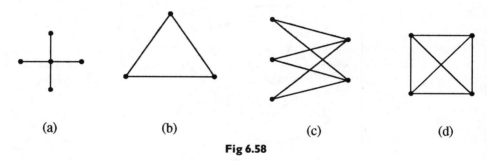

(a)  (b)  (c)  (d)

**Fig 6.58**

SOLUTION
Graphs (a) and (c) need two colours, graph (b) needs three colours and graph (d) needs four colours.

The chromatic number of a graph can be anything from 1 to $\Delta + 1$. The upper bound was proved by R. L. Brooks in 1941.

### ● *Theorem 11*

If $G$ is a simple graph whose maximum degree is $\Delta$, then $\chi(G) \leq \Delta + 1$.

PROOF
We shall use mathematical induction on $n$, the number of vertices of $G$.

*Basis of induction:*
When $n = 1$, the graph $G$ is $K_1$ for which $\chi(G) = 1$ and $\Delta = 0$. Thus, the result is true.

*Inductive step:*
Assume the result is true for all graphs on fewer than $n$ vertices. Let $G$ be a graph on $n$ vertices with maximum degree $\Delta$, and let $H$ be the graph formed from $G$ by

removing any vertex $v$. Since $H$ has fewer than $n$ vertices and maximum degree at most $\Delta$, by induction $H$ can be $(\Delta + 1)$-vertex coloured. Now, $v$ is joined to at most $\Delta$ vertices in $H$, and these vertices can have at most $\Delta$ different colours. Thus, there is at least one colour available for $v$. Hence, $\chi(G) \leq \Delta + 1$.

## • *Theorem 12*

Let $G$ be a bipartite graph, then $\chi(G) = 2$.

PROOF

Since there are no edges within either of the two partitions, we can give one colour to all the vertices in one partition and another colour to all the vertices in the other partition, and so achieve the required 2-vertex colouring.

## Map colourings

In 1852 Francis Guthrie, a geography student, noticed that a map of the counties of England could be four-coloured so that adjacent counties each had different colours. He wondered if four colours would always be sufficient for any map. He asked De Morgan, his brother's old professor. De Morgan thought it was obvious but could not prove it!

Next, De Morgan wrote to Hamilton, who thought that a proof should be possible but said that he had 'no time to do it now'. In 1878, Cayley introduced the world of mathematics to the problem by giving a seminar on **the four-colour problem** to the London Mathematical Society.

From the map shown in Fig 6.59 it is clear that **four colours are necessary**.

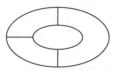

**Fig 6.59**

The problem attracted a lot of attention. Lewis Carroll, the mathematician and author of the Alice books, set the problem as a game. A headmaster set it as a challenge to his pupils saying that solutions must be less than 30 lines of writing and one page of diagrams!

In 1879 came the world's most famous 'non-proof'. Kempe, a London barrister, gave a proof which contains a fatal flaw. Nevertheless, his ideas were used as a basis for the correct proof many years later. The mistake in Kempe's proof was found by Heawood in 1890.

The four-colour theorem was finally proved by two American mathematicians, Kenneth Appel and Wolfgang Haken, in 1976. There is an account of their proof in 'The proof of the four-colour theorem' by K. Appel in *New Scientist*, 21 October 1976. The proof made extensive use of fast and large computers, and used over 1000 hours of computer time. Appel and Haken had worked on the problem for 10

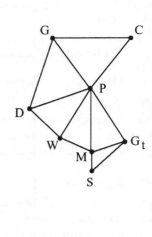

**Fig 6.60**

years. We shall not attempt a proof of the four colour theorem here! Instead, we shall prove the weaker result that six colours are sufficient.

We can convert any map to a graph by putting a vertex in each area and edges between vertices that correspond to adjacent areas. Figure 6.60 shows the map of the counties of Wales turned into a graph. Colouring the areas of the map corresponds to a vertex colouring of the associated graph.

## ● *Theorem 13*

Every connected simple planar graph is 6-vertex-colourable.

PROOF
First, we show that every connected simple planar graph always has at least one vertex with degree at most 5.

Since the graph is simple, each face has degree at least 3, so

$$2e \geq 3f$$

Combining this with Euler's Theorem we get

$$e \leq 3v - 6 \tag{1}$$

Now, if we assume that every connected simple planar graph has all vertices of degree at least 6 then

$$2e \geq 6v \tag{2}$$

Combining (1) and (2) we get $3v \leq 3v - 6$, i.e. $0 \leq -6$, which is a contradiction.

Thus, we have proved by contradiction that there is at least one vertex with degree five or less. We shall now use induction to show that any connected simple planar graph containing at least one vertex of degree at most 5 can be 6-vertex-coloured.

Let $P(n)$ be the proposition that all connected simple planar graphs on $n$ vertices can be 6-vertex-coloured.

*Basis of induction:*
Any graph with six or fewer vertices can clearly be 6-coloured.

*Inductive step:*
Let $G$ be a map with $k$ vertices. Then, $G$ contains a vertex, $v$ say, with degree at most 5. Remove this vertex. The resulting graph $G - v$ has $k - 1$ vertices and so by the induction assumption can be 6-vertex-coloured. When we reinstate the missing vertex $v$, it is joined to five or fewer colours and so there is a colour available for the vertex $v$. Hence, $G$ is 6-vertex-coloured.

## EXERCISES ON 6.8

1. True or false?
   (a) $\chi(G) \leq \Delta$.
   (b) $\chi'(G) \geq \Delta$.
   (c) $\chi(G) \leq \chi'(G)$.
   (d) It is impossible to draw a map with four countries all touching each other.
   (e) If $\chi(G) = 2$ then $G$ is bipartite.
   (f) If $\chi'(G) > 2$ then $G$ is not bipartite.
2. For each of the graphs shown in Fig 6.61 give the chromatic number and the chromatic index.

 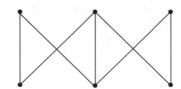 

**Fig 6.61**

3. Prove that if $G$ is a regular Hamiltonian graph of degree 3, then $\chi'(G) = 3$.
4. For the Petersen graph what are the upper and lower bounds for the chromatic index given by Vizing's Theorem? What is the exact value?
5. There are six badminton teams in the local league. Each team has to play the other five teams. A team can only play one match a week. How many weeks does it take to play all the games? If a new team is allowed to join the league so that there are seven teams in all, how many weeks does it now take to play all the games?

6.   For the Petersen graph, what is the upper bound for the chromatic number given by Theorem 11? What is the exact value?

7.   Let $G$ be a planar graph. Prove that if $G$ can be face-coloured with two colours, then $G$ is Eulerian.

8.   Prove that no map can have five mutually adjacent countries.

# 6.9 Projects

## Knight's tours

There is an old problem concerning a chessboard and the chesspiece known as the knight. In Fig 6.62, the knight can move to any one of the eight positions marked by the dots. Is it possible to find a way for the knight to visit each square of a chessboard by a sequence of knight's moves and finish on the same square as it began?

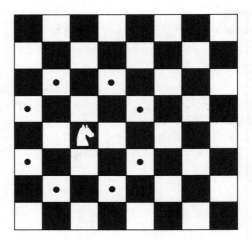

**Fig 6.62**

First, try the problem on a smaller board. Try with a $4 \times 4$ chessboard. Draw the associated graph of all possible moves, using 16 vertices to represent the squares, and edges to represent possible moves. Now see if this graph has a Hamiltonian cycle.

Next, consider a $5 \times 5$ board. If you colour the squares alternately black and white and the knight is on a black square, what possible colour square can he move to? And from a white square where can he move? Can a $5 \times 5$ board be covered by a sequence of knight's moves?

Does the $5 \times 5$ case help you to give a reason why no odd by odd board can have a knight's tour? What about a $6 \times 6$ board?

There are many solutions for an $8 \times 8$ board. Here is an interesting one which is also a magic square (the numbers in each row and column have the same sum).

| 50 | 11 | 24 | 63 | 14 | 37 | 26 | 35 |
|----|----|----|----|----|----|----|----|
| 23 | 62 | 51 | 12 | 25 | 34 | 15 | 38 |
| 10 | 49 | 64 | 21 | 40 | 13 | 36 | 27 |
| 61 | 22 | 9  | 52 | 33 | 28 | 39 | 16 |
| 48 | 7  | 60 | 1  | 20 | 41 | 54 | 29 |
| 59 | 4  | 45 | 8  | 53 | 32 | 17 | 42 |
| 6  | 47 | 2  | 57 | 44 | 19 | 30 | 55 |
| 3  | 58 | 5  | 46 | 31 | 56 | 43 | 18 |

## Travelling salesperson

A travelling salesperson wishes to visit eight cities in Europe. Below is a table giving the road distances, in units of 10 km, between each of the eight cities.

|           | Amsterdam | Berlin | Florence | Istanbul | Lisbon | Madrid | Paris | Zurich |
|-----------|-----------|--------|----------|----------|--------|--------|-------|--------|
| Amsterdam | —         | 67     | 139      | 266      | 232    | 181    | 50    | 83     |
| Berlin    | 67        | —      | 123      | 241      | 289    | 238    | 107   | 85     |
| Florence  | 139       | 123    | —        | 194      | 230    | 174    | 114   | 60     |
| Istanbul  | 266       | 241    | 194      | —        | 415    | 359    | 274   | 224    |
| Lisbon    | 232       | 289    | 230      | 415      | —      | 66     | 182   | 216    |
| Madrid    | 181       | 238    | 174      | 359      | 66     | —      | 131   | 173    |
| Paris     | 50        | 107    | 114      | 274      | 182    | 131    | —     | 64     |
| Zurich    | 83        | 85     | 60       | 224      | 216    | 173    | 64    | —      |

If we calculated all possible Hamilton cycles there would be 8! = 40 320 of them. Clearly this is too many to do by hand.

This project will look at a way of getting upper and lower bounds for a minimum length cycle. Choose any possible route and calculate the length. This is an upper bound. Why? Can you find a better upper bound?

To find a lower bound, suppose we have found a minimum length cycle. Then, if we remove a vertex, $v$ say, and its two associated edges, we will have a minimum spanning tree of the graph without vertex $v$. Why?

The length of the minimum cycle is the length of this spanning tree plus the length of the two edges which we removed from $v$ (Fig 6.63).

We can find a lower bound for the travelling salesperson by taking our original graph, $G$, removing a vertex, say $v$, finding a minimum spanning tree in $G - v$ and finally adding the two shortest-length edges from $v$.

Try this for the eight cities. First remove Amsterdam. What lower bound does this give? For each city in turn, repeat the process until you have eight lower bounds. Which is the best lower bound?

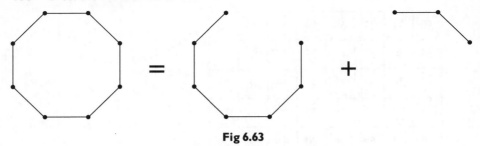

**Fig 6.63**

## Regular convex polyhedra

We can use Euler's formula to find out about regular convex polyhedra. Figure 6.64 shows the graphs of the tetrahedron and the cube obtained by imagining that the corners of the polyhedron touch the surface of a sphere and then projecting the polyhedron onto a planar graph.

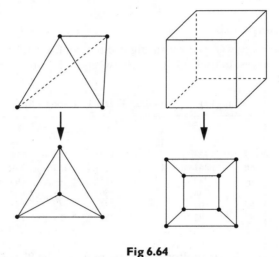

**Fig 6.64**

A regular convex polyhedron is equivalent to a regular graph in which each face has the same number of sides, all vertex degrees are at least 3, and all faces have at least three sides.

Let $v$, $e$ and $f$ be the numbers of vertices, edges and faces, let $d$ be the degree of the vertices, and $g$ the number of edges around each face.

Obtain a formula for $v$ in terms of $e$ and $d$, and a formula for $f$ in terms of $e$ and $g$.

Using Euler's formula, and replacing $v$ and $f$ with the expressions you have found, you can show that

$$\frac{1}{d} - \frac{1}{2} + \frac{1}{g} = \frac{1}{e}$$

Since $1/e > 0$ it follows that

$$\frac{1}{d} + \frac{1}{g} > \frac{1}{2}$$

We also know that $d \geq 3$ and $g \geq 3$.

If we try $d = 3$ and $g = 3$ then our formula for $1/e$ gives

$$\frac{1}{3} - \frac{1}{2} + \frac{1}{3} = \frac{1}{e}$$

so $e = 6$. Using the formulae you obtained for $v$ and $f$, you can show that $v = 4$ and $f = 4$. This corresponds to the graph of the tetrahedron.

By trying different values for $d$ and $g$, you can show that there are only five possible graphs. These graphs correspond to the five Platonic solids: tetrahedron, cube, octahedron, dodecahedron and icosahedron.

## Instant insanity

This is a puzzle which consists of four cubes whose sides have been painted with the colours blue, yellow, green and white. The idea is to see if the cubes can be piled up into a $4 \times 1$ tower so that all four colours appear on each of the four sides. Fig 6.65 shows the diagrams for the four cubes.

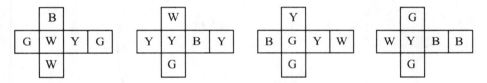

**Fig 6.65**

Either make up some cubes from cardboard and colour them as shown, or borrow four wooden cubes from an obliging child and label the sides with the respective colours. See if you can arrange the cubes into the required tower. By trial and error it can take a very long time! How many ways are there to pile up the four cubes into a $4 \times 1$ tower?

We shall represent each cube as a graph and see how this helps us find a solution.

For each cube, we have four vertices labelled blue, yellow, green and white. We draw an edge between two vertices if the two colours are on opposite faces of the cube.

Fig 6.66 shows the four graphs for the four cubes.

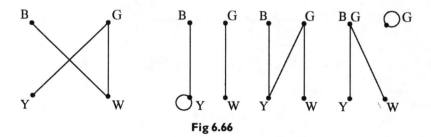

**Fig 6.66**

We can superimpose the four graphs on one, labelling the edges to show which cube they are from (Fig 6.67).

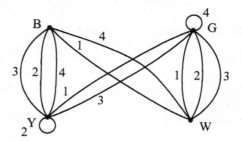

**Fig 6.67**

We now find two independent subgraphs in Fig 6.67 (that is, the two subgraphs do not have any edges in common), where each subgraph contains exactly one edge from each graph and is regular of degree two (Fig 6.68).

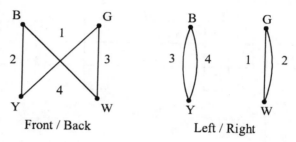

**Fig 6.68**

One subgraph gives us the front and back positions for each cube, whilst the other gives the left and right positions. Use these two graphs to help you position the four cubes (Fig 6.69).

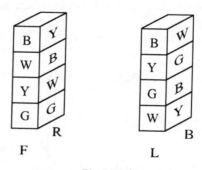

**Fig 6.69**

Fig 6.70 shows two more sets of cubes for you to try. Find a different set of four cubes. Will all possible colourings be solvable?

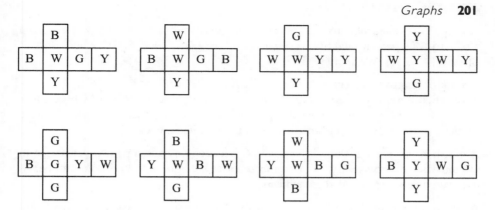

**Fig 6.70**

# *Summary*

A **graph** $G$ consists of a non-empty set $V(G)$ of **vertices** and a list $E(G)$ of unordered pairs of these elements called **edges**. If $v$ and $w$ are two vertices of $G$ then an edge of the form $vw$ is said to **join** $v$ and $w$. Let $e(G)$, $v(G)$ denote the numbers of edges and vertices of $G$.

Two or more edges joining the same pair of vertices are called **multiple edges**. An edge joining a vertex to itself is called a **loop**. A graph with no loops or multiple edges is called a **simple graph**.

A **subgraph** of a graph $G$ is a graph all of whose vertices belong to $v(G)$.

The **degree** of a vertex $v$ of $G$ is the number of edges incident with $v$ and is denoted by $d(v)$. We denote by $\delta(G)$ and $\Delta(G)$ the minimum and maximum degrees, respectively, of vertices of $G$. We can write the degrees down in non-decreasing order to obtain the **degree sequence** for $G$.

**Complete graphs** are simple graphs in which each pair of distinct vertices is joined by an edge. The complete graph on $n$ vertices is denoted by $K_n$.

**Bipartite graphs** are graphs whose vertex set can be partitioned into two subsets $X$ and $Y$ so that each edge has one end in $X$ and one end in $Y$ and there are no edges between the vertices of $X$ or between the vertices of $Y$. **Complete bipartite graphs** are simple bipartite graphs with bipartition $(X, Y)$ in which each vertex of $X$ is joined to each vertex of $Y$. The complete bipartite graph on $m$ vertices in $X$ and $n$ vertices in $Y$ is denoted by $K_{m,n}$.

The **complement** of a graph $G$ has the same set of vertices as $G$, and its edges are the edges that are missing in $G$. The complement of $G$ is denoted by $\bar{G}$.

The **adjacency matrix** $A(G)$ is the $n \times n$ matrix in which $A_{ij}$ is the number of edges joining $v_i$ and $v_j$. The **incidence matrix** $I(G)$ is the $n \times m$ matrix in which $I_{ij}$ is the number of times that $v_i$ and $e_j$ are incident.

A graph $G$ is **isomorphic** to a graph $H$ if there exists a one-to-one mapping $\phi$ from $V(G)$ to $V(H)$ such that $\phi$ preserves adjacency, i.e. $uv \in E(G)$ if and only if $\phi(u)\phi(v) \in E(H)$. We write $G \cong H$.

A **walk** of length $k$ is a succession of $k$ edges of $G$. A walk of length $k$ all of whose edges are different is called a **trail** of length $k$. A walk of length $k$ all of whose edges *and* vertices are different is called a **path** of length $k$. A closed walk all of whose vertices are different (except the first and last) is called a **cycle**.

A graph is **connected** if there is a path in $G$ between any given pair of vertices, and **disconnected** otherwise. Every disconnected graph can be split up into a number of connected subgraphs called **components**.

A **tree** is a connected graph with no cycles. A graph all of whose components are trees is called a **forest**.

A **spanning tree** is a subgraph of $G$ that includes every vertex of $G$ and is a tree. If each edge of the graph has a length (or cost) associated with it then a **minimum spanning tree** is a spanning tree of minimum length.

A trail that visits every edge of the graph exactly once is called an **Eulerian trail**. A connected graph $G$ is **Eulerian** if there is an Eulerian trail. A graph with a trail that covers every edge exactly once but need not begin and end at the same place is called **edge-traceable.**

A path that contains every vertex of $G$ exactly once is called a **Hamiltonian path** of $G$; similarly a **Hamiltonian cycle** of $G$ is a cycle that contains every vertex of $G$ exactly once.

A graph is said to be **planar** if it can be drawn in the plane in such a way that no two edges meet each other except at a vertex. If $G$ is a planar graph, then any plane drawing of $G$ divides the plane up into regions called **faces**. The **degree of a face** is the number of edges around it, and is denoted by $d(f)$. We also consider the outside region to be a face.

If we insert vertices on the edges of a graph we obtain a **subdivision** of the original graph.

A **$k$-edge colouring** of a (loopless) graph $G$ is an assignment of $k$ colours to the edges of $G$ in such a way that any two edges meeting at a common vertex have different colours. The **chromatic index** of $G$, denoted by $\chi'(G)$, is the smallest number $k$ for which $G$ has a $k$-edge colouring.

A **$k$-vertex colouring** of a loopless graph $G$ is an assignment of $k$ colours to the vertices of $G$ such that adjacent vertices receive different colours. The **chromatic number** of $G$, denoted by $\chi(G)$, is the smallest number $k$ for which $G$ has a $k$-vertex colouring.

# Glossary

| | |
|---|---|
| $\forall$ | for all |
| **and** | the logical operator and |
| $\binom{n}{k}$ | combination of $n$ from $k$ |
| $\equiv$ | congruent |
| $\delta$ | minimum degree |
| $\Delta$ | maximum degree |
| $d(v)$ | degree of $v$ |
| $\setminus$ | difference |
| $\mid$ | divides |
| $e(G)$ | number of edges of $G$ |
| $\emptyset$ | empty set |
| $\exists$ | there exists |
| $\Leftrightarrow$ | if and only if |
| $\Rightarrow$ | implies |
| $\in$ | is an element of |
| $\lfloor x \rfloor$ | greatest integer less than $x$ |
| $\bar{G}$ | complement of $G$ |
| $\cap$ | intersection |
| $\cong$ | isomorphic |
| $f^{-1}$ | inverse of the function $f$ |
| $K_n$ | the complete graph on $n$ vertices |
| $K_{m,n}$ | the complete bipartite graph on $m$ and $n$ |
| **not** | the logical operator not |
| **or** | the logical operator or |
| $P(\cdot)$ | probability of |
| $\mathcal{P}(X)$ | power set of $X$ |
| $\mathbb{N}$ | natural numbers |
| $\mathbb{Z}$ | integers |
| $\mathbb{Q}$ | rational numbers |
| $\mathbb{Q}^+$ | positive rational numbers |
| $\mathbb{R}$ | real numbers |
| $\mathbb{R}^+$ | positive real numbers |
| $\mathbf{R}$ | is related to |
| $\subseteq$ | is a subset of |
| $|X|$ | number of elements in $X$ |
| $\chi(G)$ | chromatic number |
| $\chi'(G)$ | chromatic index |
| $\cup$ | union |
| $\mathcal{U}$ | universal set |
| $v(G)$ | number of vertices of the graph $G$ |

# Index